HUMAN ANATOMY AND PHYSIOLOGY

THE BIG SLIDE

THE DIGESTIVE SYSTEM FROM TOP TO BOTTOM

Human Anatomy and Physiology

Additional books in this series can be found on Nova's website under the Series tab.

Additional e-books in this series can be found on Nova's website under the e-book tab.

HUMAN ANATOMY AND PHYSIOLOGY

THE BIG SLIDE

THE DIGESTIVE SYSTEM FROM TOP TO BOTTOM

NICHOLAS F. PRAYSON
AND
RICHARD A. PRAYSON

Illustrator: J. Jordi Rowe

NOVA BIOMEDICAL

New York

Library of Congress Cataloging-in-Publication Data

ISBN: 978-1-63482-840-6

Library of Congress Control Number: 2015939087

Published by Nova Science Publishers, Inc. † New York

With love, the authors and illustrator dedicate this book to
Beth, David, Brigid, Iain, Cameron and Diana

Contents

Preface

The human body is tremendously complex in its workings. There is a lot we have come to understand about it over the centuries, and yet, there is so much we do not understand or we misunderstand. A hundred years from now, scientists will look back at today's "truths" and will likely wonder how we could have been so ignorant.

This text attempts to present an overview of the human digestive system, breaking it down into its individual components, examining what we currently know or understand. You do not have to be a scientist to read this book – the book is not written for scientists. It is written for those who want to learn a little more about the human body and how it works. The book presents the basics of anatomy (how various parts of the digestive system look, where the parts are located and how they are organized), physiology (how they function), and pathophysiology (how they sometimes do not function as they should). Rather than organize the book like a traditional textbook, a question and answer format was adopted to help organize each of the chapter's content. We have also dug up some interesting trivia and history related to many of the parts of the digestive system that are discussed; hopefully, the reader will find these entertaining.

So climb to the top of the slide and get ready to take a trip down through the human digestive system.

Chapter 1

The Big Slide: Overview

Have you ever wondered what happens to the apple or piece of cake that you eat? It may seem obvious at first glance. The food that you eat provides fuel for your body to work properly and to keep you healthy. We all need to eat food to refuel our bodies. The food provides nutrients that give different parts of the body the energy they need to work. But how does your body take an apple or piece of cake and change them into nutrients that make your arm move or your brain think?

The parts of your body that are responsible for transforming the food that you eat to nutrients are part of the digestive system. The digestive system is similar to a big slide. The digestive system can be viewed as a series of tubes that start at the mouth, where the food goes in, and extends all the way down to the anus, where the feces or stool exits (Figure 1.1). The trip typically takes a day or two.

Along the way, the food is ground up and broken down into smaller and smaller particles. Eventually, these small particles or nutrients are absorbed or taken in by the digestive system and then distributed throughout the rest of the body. Digestion is the process by which food is broken down into its smallest parts so that the body can use them to function.

Now let us start the slide through the digestive system at the very top with the teeth and mouth.

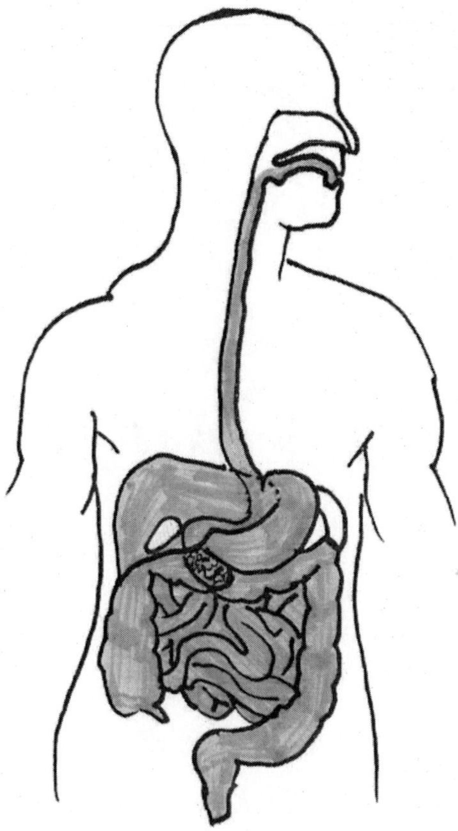

Figure 1.1. This schematic drawing outlines the location of the digestive tract within the body.

Chapter 2

Teeth

Digestion of food begins in the mouth. The teeth play an important role in this process by mechanically breaking down the food that we eat into smaller pieces (Figure 2.1).

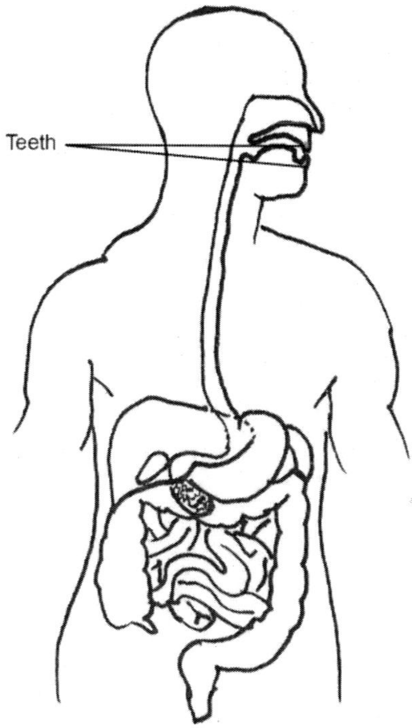

Figure 2.1. This is a schematic drawing of the digestive tract showing where the teeth are located.

What Is the Difference between Baby Teeth and Adult Teeth?

Most babies are born toothless. Teeth usually do not begin to appear until a baby is 6 to 12 months old. These early teeth are called baby or primary teeth. Most baby teeth are present by 3 years of age. There are 20 baby teeth in all. At about age 5 or 6 years, these teeth begin to fall out. They actually fall out because they are being forced out by the larger adult or permanent teeth which are pushing their way out from beneath the baby teeth.. By age 12 or 13 years, all the baby teeth are gone and have been replaced by the adult teeth. There are 28 adult teeth. Four additional adult teeth appear a bit later, usually between ages 16-21 years. These late arriving teeth are located way in the back and are called wisdom teeth (think, older equals wiser!). After they arrive, this brings the total number of teeth in an adult to 32.

What Are the Different Types of Teeth and What Do They Do?

There are several different types of teeth, whose shapes differ from one another due to the function that they serve (Figure 2.2). The teeth in the very front are called incisors. There are 4 of them on the top and 4 on the bottom. The incisors are shaped like chisels with flat ends that are sharp. They are designed to cut food (think of biting into an apple) and shovel the food inward.

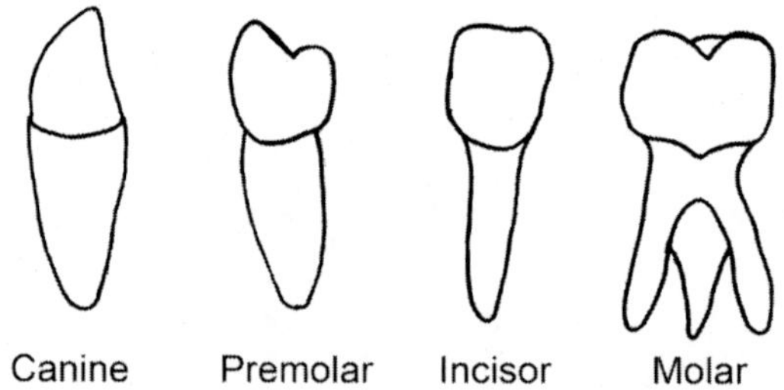

Figure 2.2. This schematic illustrates the various shapes of the teeth. The shape is suited to the function. The incisors, which are located in the front of the mouth, have a sharp edge at the surface that is useful for cutting food. The canines have a sharp pointy surface useful for grasping and tearing food. Premolars have a flattened surface useful for crushing and grinding food. The molars, which are situated in the back of the mouth, have a flat surface with ridges which are useful for chewing and grinding food.

Next to the incisors are pointy teeth called canines. There are 2 on the top and 2 on the bottom. Their pointy shape and sharpness are useful in grasping and tearing food.

Premolars are located next to the canines. They are also called bicuspid teeth. There are 4 premolars on the top and 4 on the bottom, 8 in all. Premolars are bigger teeth with a flattened surface. Their major job is to crush and grind food.

Behind the premolars are the molars. They are even larger than the premolar teeth and have a somewhat flat surface with ridges. There are also 8 molars (4 on the top and 4 on the bottom). Their job is to both chew and grind food and they help in swallowing the food.

The last teeth to arrive are all the way in the back. These are the 4 wisdom teeth or third molars (two on the top and two on the bottom). They do not serve any particular job. Sometimes when they do not come in properly, they can cause problems and need to be removed.

What Makes Teeth So Hard?

The part of the tooth that you normally see when you open your mouth is known as the crown. The crown is covered by a tough outer coating that is called the enamel. The enamel serves as a protection for the tooth against the wear and tear of chewing. Enamel is actually the hardest substance in the entire body and is one of the main reasons why teeth are so hard. Underneath the enamel is the dentin. Dentin makes up most of the substance of the tooth and it provides a second layer of protection for the innermost part of the tooth, the pulp. The nerves and blood supply to the tooth are located in the pulp. The nerves are important in detecting the pain. They are responsible for the pain of a tooth ache but more on that later. The blood supply in the pulp brings nutrients to the tooth to keep it alive. All of these structures which are located above the surface of the gum line are considered to be part of the crown of the tooth. Below the surface of the gum line, the root of the tooth extends down toward the jaw bones. About 2/3 of the total length of the tooth is made up of root. Much of the root is made up of cementum, which helps in anchoring the root to the jaw bone.

Besides Chewing and Grinding Food, What Else Do Teeth Do?

Teeth are also important in talking. Your teeth work together with the tongue in order to make different sounds. Say the work "tooth" aloud. When you pronounce the "t", notice how your tongue touches the sides of the incisors. The tongue then moves between the upper and lower front teeth to make the "th" sound. Beyond helping you speak, your teeth are also an important part of your appearance, especially when you smile!

What Is a Cavity and Why Do They Sometime Hurt?

A cavity is a hole that develops in teeth. This is due to bacteria that are present in large numbers inside the mouth. The bacteria grow on the surface of the tooth enamel. This bacteria gets mixed with saliva or spit which can form a white film on the surface of the tooth called a plaque. The bacteria that continue to grow in the plaque use sugar for energy. Some bacteria can also use sugar to make a glue-like substance that helps make them stick harder to the tooth enamel. Acid that is produced by the bacteria in the plaque eat away at the enamel, eventually forming a hole or cavity (Figure 2.3). If the cavity becomes deep enough, it irritates the nerve fibers in the pulp area, causing a toothache.

How Does Brushing Your Teeth Prevent Cavities?

The toothbrush is important in removing plaque from the surface of the tooth. Brushing with a soft-bristle toothbrush works well. Harder-bristle toothbrushes or hard brushing can actually begin to wear away tooth enamel. This can eventually make the tooth more likely to develop cavities. How does toothpaste help the cleaning process? Toothpaste contains a chemical called fluoride which helps to make the enamel more durable. Dental floss is also useful in removing plaque from between the teeth, where the toothbrush has a hard time reaching.

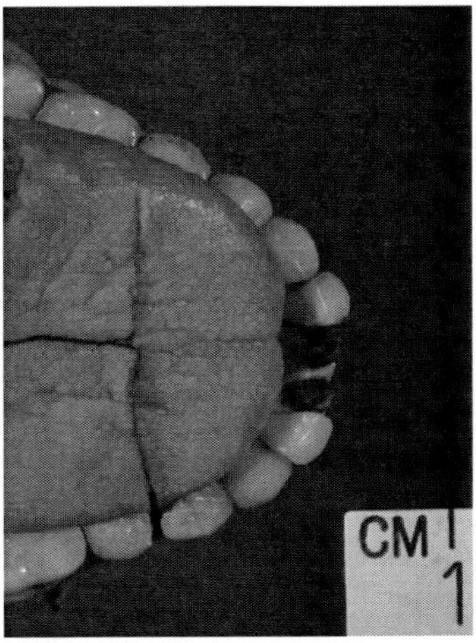

Figure 2.3. This is a picture of the front part of the tongue and teeth. Notice the black areas in the two front teeth; these are cavities.

Why Do a Lot of Kids Wear Braces and Retainers?

Straight teeth are important to have. Crooked teeth can cause several problems: 1) they can affect the way a person chews; 2) they can affect speech; 3) they can create areas that are tough to clean with toothpaste and proper brushing; 4) they can cause the jaw to line up incorrectly, which can eventually lead to jaw pain; and 5) they can affect the way a person looks. Braces and retainers can be used to help correct crooked teeth and straighten them. The braces are attached directly to the teeth. Gentle pressure is applied to the teeth by adjusting the braces. Over time, the braces can slowly move the teeth into their proper position. A retainer is often used for less severe problems or after braces in order to continue to keep the teeth in proper alignment. Retainers can be easily put in and removed by the user.

Interesting Facts

1) Most babies are born without teeth. One out of 2,000 babies is born with a tooth. (www.mountainviewfamilydentistry.com/ourblog/?p=332)

2) Cats have 30 teeth. Dogs have 42 teeth. Armadillos have 104 teeth (www.funanimaldentalfacts/roscoevillageanimalhospital.com/fun-dental-facts)
3) In the United States, about 3 million miles of dental floss is bought in a year.
(www.carabelli.com/pages/patient_library/trivia.htm)
4) Elephant tusks are a special type of incisor which is used for digging up food, lifting things and fighting. The longest tusk ever recorded was 138 inches long and weighed 314 pounds.
(www.liveanimalslist.com/mammals/elephant-tusk-facts)
5) One old fashioned cure or prevention for toothaches involved boiling earthworms in oil and then using the oil as eardrops. As a cure for loose teeth, a frog was tied to the jaw.
(www.carabelli.com/pages/patient_library/trivia.htm)
6) Tooth decay is the second most common disease in the United States after the common cold.
(www.carabelli.com/pages/patient_library/trivia.htm)
7) During the Dark Ages, it was believed by some that a person could cure her or his toothache or regrow a lost tooth by merely possessing a tooth from another person, especially a hanged criminal. (www.carabelli.com/pages/patient_library/trivia.htm)
8) The earliest recorded use of toothpaste was in 1700s. One of the main ingredients used for scrubbing the teeth was burnt bread. (en.wikipedia.org/wiki/toothpaste)
9) The strongest teeth in the world belonged to John Massis of Belgium. He pulled 15 tons of train along a railroad track with a rope on a bit between his teeth 15 meters (www.legendarystrength.com/john-massas)

Chapter 3

Oral Cavity

Besides the teeth, which begin the process of digestion by breaking down food into smaller pieces, there are other structures in and around the mouth that are important in the process of digestion.

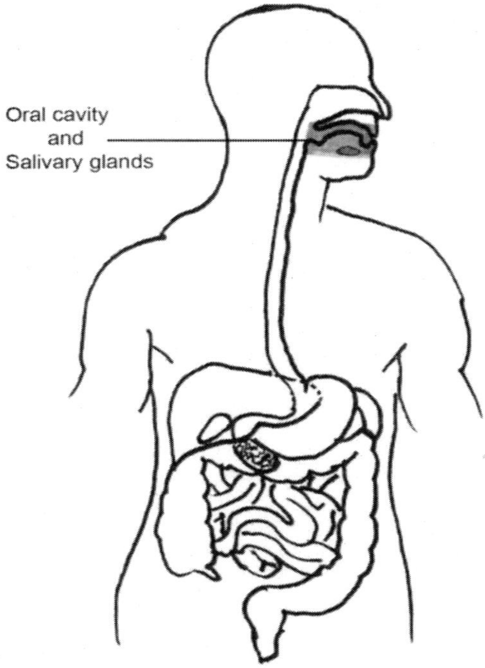

Figure 3.1. This is a schematic drawing of the digestive tract showing where the oral cavity and the salivary glands are located.

These structures include the tongue, mouth and salivary glands (Figure 3.1). The mouth is lined by a special layer of cells called squamous epithelium that helps provide a protective barrier against infection (Figure 3.2). The mouth also contains others structures which help fight infection, such as the tonsils, which structures that are made up of many small inflammation fighting cells known as lymphocytes (Figure 3.3). The tonsils are located in the back of the mouth area in the throat. Sometimes, they get infected and swollen and need to be removed.

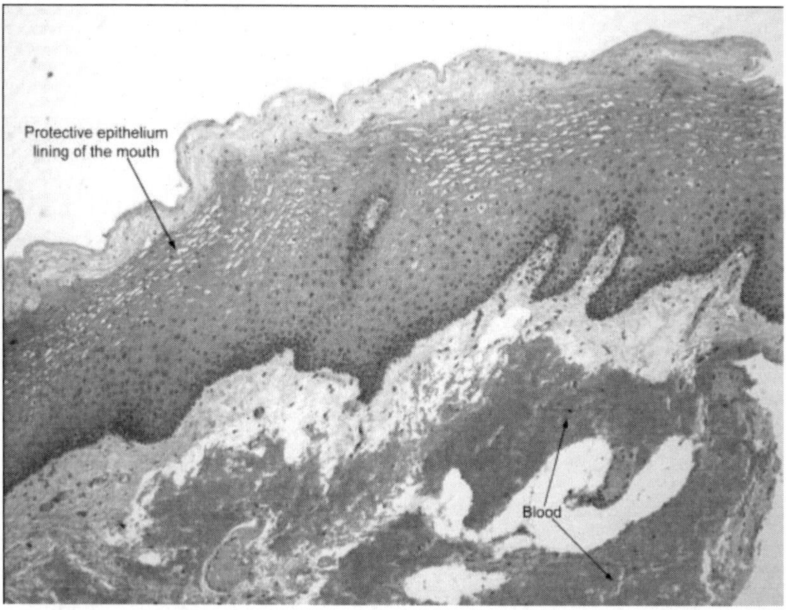

Figure 3.2. This is a picture of tissue taken from the mouth and viewed under a microscope, magnified 400 times. The mouth is lined by a special layer of cells, called squamous epithelium, which helps provide a barrier against infection. The bright red areas at the bottom represent blood. The person whose mouth this was had a problem which caused some bleeding in the mouth.

What Does Saliva Do and Where Does It Come from?

Saliva or spit is a clear liquid which is manufactured by structures known as the salivary glands (Figure 3.4). The biggest salivary gland is the parotid gland. It is located toward the back of each cheek between the ear and jaw.

There are other smaller salivary glands (sublingual and submandibular glands) located in different places in the mouth. When you eat something, a signal from the brain tells the salivary gland to release saliva into the mouth (Figure 3.5).

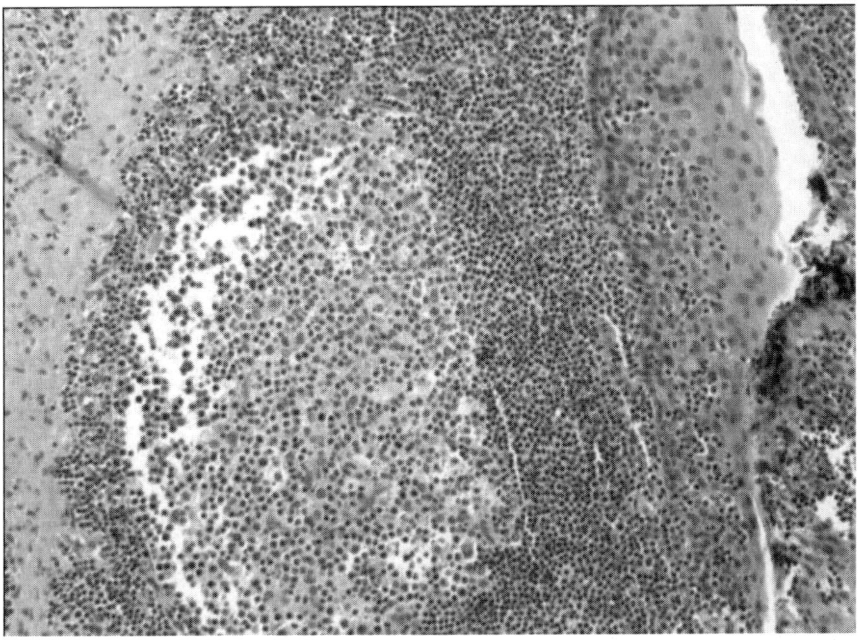

Figure 3.3. This is a picture of what a tonsil looks like under the microscope, magnified 200 times. The tonsil is made up of many dark staining cells called lymphocytes that help fight against infection.

Saliva serves two major purposes. It allows the food to be swallowed more easily by mixing with it, making it mushy. Saliva is also important in the process of breaking down food. Saliva contains a substance known as salivary amylase. Salivary amylase is a protein in the saliva that breaks apart starches in foods such as fruits, breads and cereals into simpler sugars that the body can use for energy. If you chew on these foods long enough, they will begin to taste sweet. This is because the amylase is breaking the starch down into sugars and you are tasting the sugar!

In addition, saliva also contains another protein, called lysosome, that plays a role in killing bacteria. Bacteria enter the mouth along with food. The saliva helps the body prevent infection from the bacteria by attacking and killing it while it is still in the mouth.

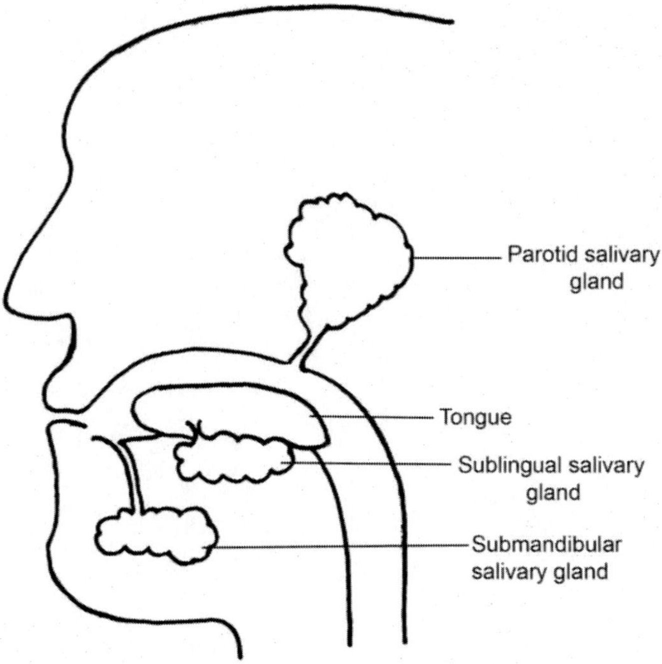

Figure 3.4. This is a schematic drawing showing where the three main salivary glands (parotid, sublingual and submandibular glands) are located and where saliva from the glands is dumped into the oral cavity.

What Is Mumps?

Mumps is an infection of the salivary glands caused by a virus. Viruses are tiny organisms that attack various parts of the body and make you sick. Specifically, the mumps virus enters cells in the salivary gland and uses the cell's machinery to manufacture more of itself; these additional viral organisms can go and infect other cells. The flu is an example of another illness caused by a virus. When the mumps virus attacks the salivary glands, it causes the gland to get larger or swell. The swelling causes the gland to be painful. Because mumps was so common in the past, most children receive a vaccination to help prevent getting the virus. Many children now receive the vaccine between one and one and a half years of age. They then receive a second vaccine dose between the ages of 4 and 6 years.

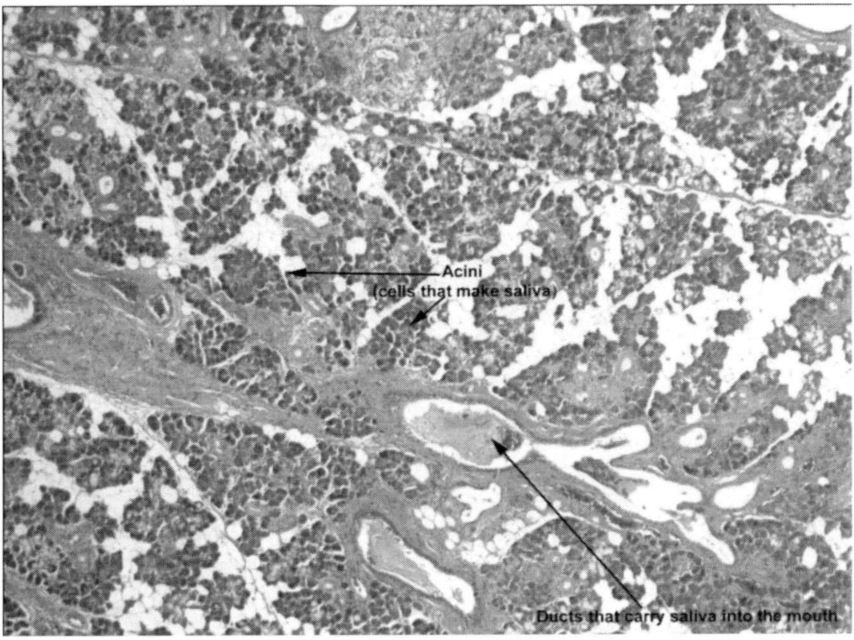

Figure 3.5. This is a picture of what a salivary gland looks like under the microscope, magnified 50 times. The darker cells make the saliva. The saliva is carried out of the salivary gland via ducts into the mouth.

What Allows Us to Taste Food?

Imagine what it would be like to not be able to taste the food that we eat. Food cannot be tasted unless it is mixed with saliva. But what part of the body is responsible for tasting food? It is mostly the tongue. If you look in the mirror while sticking out your tongue you will see that the surface of the tongue is covered by tiny bumps called papillae (Figure 3.6). Tiny taste organs called taste buds are found in the papillae. Many taste buds are found on the surface of the tongue, but they may also be found under the tongue, on the sides of the cheeks (inside the mouth) and on the roof of the mouth. When the taste buds come in contact with food or drink, they detect the taste and send a message to the brain which allows it to interpret the taste as sweet, sour, salty or bitter. Different regions of the tongue are primarily responsible for detecting different types of tastes (Figure 3.7). For example, the taste buds on the back of the tongue are responsible for detecting bitter tasting foods, while those on the front of the tongue detect sweet foods.

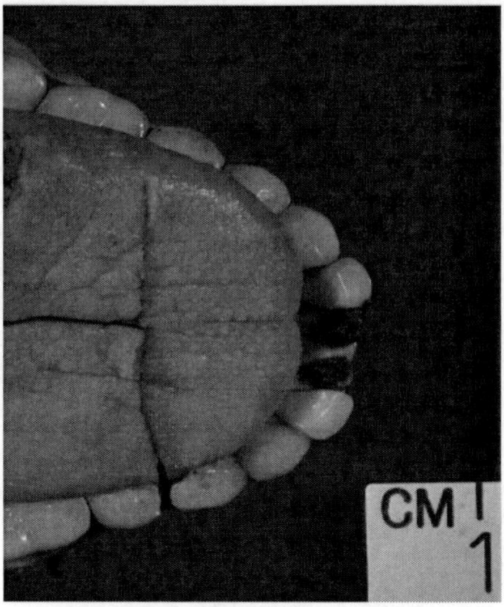

Figure 3.6. This is the same picture as in Chapter 2 (Figure 2.3). Look carefully at the surface of the tongue and notice the small bumps. These bumps are known as papillae; these are where the taste buds are located.

Why Do Foods Taste Funny When You Have a Cold?

There are other things besides taste buds that can affect our ability to taste things. Did you ever have a cold and notice that food is harder to taste? Smelling also plays a large role in our ability to taste things. Temperature can also influence the way things taste. If you warm the front part of the tongue, a sweet sensation can be generated. Cooling the tongue leads to a sour or salty taste.

Why Don't We Swallow Our Tongue?

The tongue also plays a role in moving the food to the back of the mouth to be properly chewed and eventually swallowed. The tongue is made up of many groups of muscles. The muscles are arranged in different directions. This allows you to move your tongue in many different ways. If you look again in a mirror under your tongue, there is a membrane that connects your tongue to the bottom of the mouth. This membrane is called a frenulum. The frenulum prevents you from swallowing your own tongue.

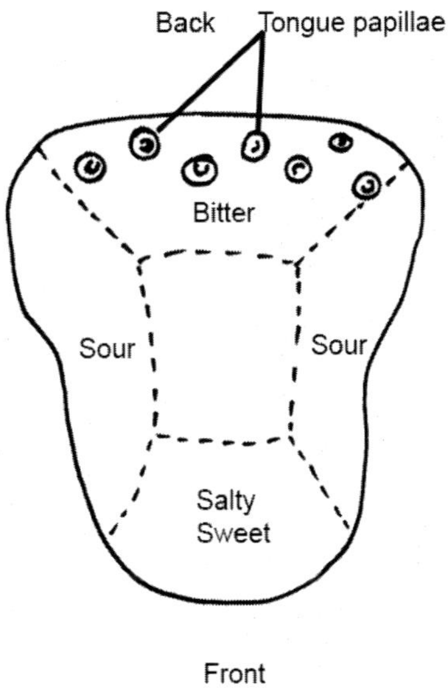

Figure 3.7. This schematic drawing shows where various types of tastes are sensed by the tongue's taste buds.

How Does the Tongue Help with Talking?

The tongue is important in speech. The muscles in the back part of the tongue are used in making certain sounds. Slowly try and make a "k" sound (like in the word "kick") or a hard "g" sound (like in the word "dog"). Notice how the back of the tongue moves against the top of the mouth to help create the sound.

Interesting Facts

1) The average human makes about one quart of saliva a day; that is 10,000 gallons in a lifetime. (www.strange-facts.info/interesting-facts-about-saliva)

2) Several hundred years ago, it was believed that cow saliva could be used to treat male baldness. (www.keratin.com/ay/quack-cures-for-hair-loss.shtml)
3) Cats lose almost as much fluid in the saliva while grooming themselves as they do through urination or peeing. (www.fun factz.com/tags/saliva)
4) The average horse produces 10 gallons of saliva each day. (www.aranchhorse.com/10-facts-about-horses)
5) The mouth is the most unsanitary part of your whole body. An estimated 20 billion bacteria live there. (www.rdhmag.com/.../oral-bacteria-how-many-how-fast.html)
6) The reason most mosquito bites itch is that mosquitos inject saliva into the bite site first before sucking out your blood. The mosquito saliva numbs the area, while the mosquito sucks out your blood, but later, it causes itching. (www.wisegeek.org/why-do-mosquito-bites-itch.html)
7) Girls generally have more taste buds than boys. (www.understanding-life.org/ask-a-physiologist/it-true)
8) An average of 2000-8000 taste buds are located on the tongue. Certain things like smoking, alcohol use, caffeine use, spicy food and hot food can dull the taste buds. The number of taste buds each person has decreases with age. The average lifespan of a taste bud in about 10 days. (en.wikopedia.org/wiki/Taste-bud)
9) The longest human tongue ever recorded belonged to Stephen Taylor; it measured 3.74 inches. (www.worldslongesttongue.com)
10) Some animals like frogs, lizards and anteaters also use their tongues to gather food. The tongue of an anteater may be up to 2 feet long. Its tongue is also covered with sticky saliva in order to help pick up bugs. (www.animals.sandiegozoo.org/animals/giant-anteater)
11) Snakes have a forked tongue, which they use to sense their surroundings. In a way, when a snake is flicking its forked tongue, it is just smelling the surrounding environment. The snake uses its tongue to collect particles from the air which it brings into its mouth to analyze. (www.petmd.com/reptile/pet-lover/evr-rp-why-snakes-use-tongue)
12) You can tell that a gorilla is angry when it sticks out its tongue. (www.funnyfact.com/gorillas-do-this-when-they-are-angry)

Chapter 4

Esophagus

Where Does the Food Go When It Is Swallowed?

The tongue and the muscles in the mouth push the chewed food, which is now mixed with saliva, down the throat toward the epiglottis. The epiglottis is a flap in the throat which closes over the opening to the airways so that the food does not get into the airway and prevent you from breathing! Sometimes the epiglottis does not cover the airway in time. This causes a severe fit of coughing in order to force the food out of the airway.

What Is the Esophagus?

The esophagus is shaped like a long straight tube which measures about 25 cm or 10 inches in length in an adult (Figure 4.1). After being swallowed, the food enters the esophagus. The inside surface of the esophagus is smooth in appearance and is lined by a layer of squamous epithelium, just like the mouth (Figure 4.2). How does the food move down the esophagus? If you guessed gravity, you are wrong. The food is pushed along its way down the esophagus by muscles (called smooth muscle) that make up part of the wall of the esophagus. These muscles contract in waves which squeezes the food down the esophagus. This process is called peristalsis (Figure 4.3). It is important to recognize that the body contains different types of muscle. The smooth muscle in the wall of the esophagus is different than the skeletal muscle that makes up the muscles in your arms and legs and cardiac muscle which makes up most of the heart.

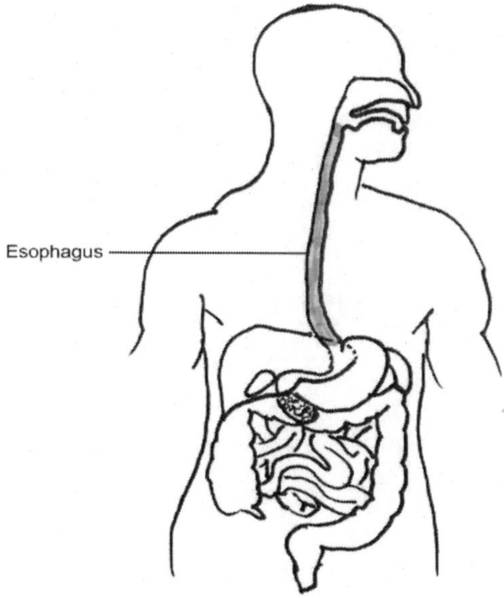

Figure 4.1. This schematic drawing shows where the esophagus is located in relation to the body and the rest of the digestive system.

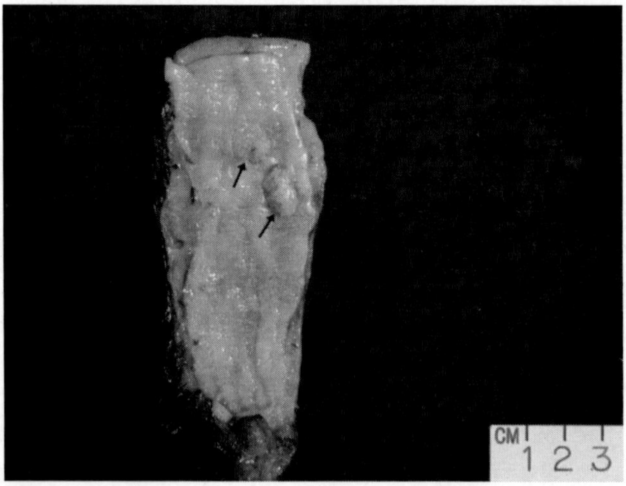

Figure 4.2. This picture shows what the inside of the esophagus looks like. This person has two small elevated areas (arrows) on the inside which represent a cancer.

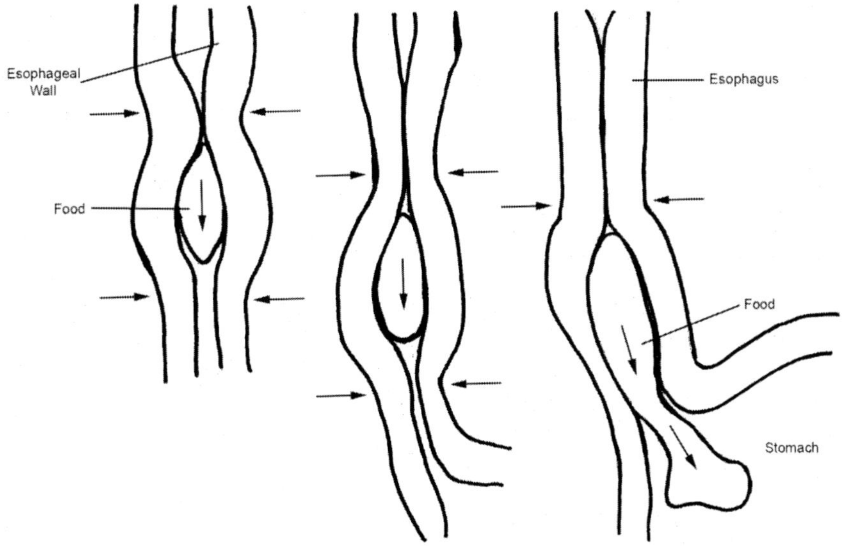

Figure 4.3. This schematic drawing illustrates what happens in peristalsis. With peristalsis, food is moved downward in the esophagus toward the stomach due to rhythmic contractions of the walls of the esophagus.

Where Does the Food Go after It Reaches the Bottom of the Esophagus?

When the food reaches the bottom of the esophagus, it goes through a thick muscular ring called a sphincter in order to enter into the stomach. The sphincter prevents the food from leaving the stomach and coming back up into the esophagus.

What Causes Vomiting?

Vomiting occurs when food and liquid comes back up from the stomach through the esophagus and out of the mouth. There are many things that can cause vomiting. Eating food that has a lot of bacteria is a common cause of vomiting. Ingesting or drinking too much of certain drugs and toxic substances, such as alcohol, can also cause vomiting. The formal medical term that doctors use for vomiting is "emesis".

What Causes Heartburn?

Heartburn, which is also known as gastric reflux disease, is a feeling of burning or pain in the center of the chest. This is caused by acid in the stomach splashing back up into the esophagus. The acid irritates the inside lining of the esophagus (the squamous epithelium) and causes the pain. There is very little acid normally in the esophagus but lots of it in the stomach, where it aids in the digestion of food. Certain things such as eating too much food or drinking liquid with lots of caffeine, like coffee, can cause heartburn to occur.

Interesting Facts

1) The esophagus of an ostrich is 1.2 meters long and is twisted. (www.cnaa.acad.md/en/thesis/6828)
2) Certain birds use their esophagus to also make noises. (www.biology.allaboutbirds.org/tag/esophagus)
3) It takes food about 7 seconds to go from the mouth down the esophagus into the stomach. (www.wisegeek.com/how-long-does-it-take-food-to-reach-your-stomach)
4) In an octopus, the esophagus goes right through the brain. (www.octopusworlds.com/octopus-anatomy)

Chapter 5

Stomach

Where Is the Stomach Located?

The stomach is attached to the end of the esophagus. It is located below the diaphragm, which is a thin muscle which separates the chest cavity (also known as the thorax) where the heart and lungs are located) from the abdominal cavity (where the liver and intestines are located). The stomach is contained within the abdominal cavity (Figure 5.1).

What Does the Stomach Look Like?

The stomach is a stretchy sac that is shaped like a large bean. The long side of the bean is called the greater curvature and the shorter side is called the lesser curvature. The stomach itself is divided into four regions (Figure 5.2). The food coming down from the esophagus enters the fundus, which is formed by the upper greater curvature of the stomach and corpus (which is Latin for body) of the stomach. From the fundus, the food migrates down into the corpus, and then proceeds to the antrum., which comprise the lower part of the stomach. When the stomach has completed its job of breaking down the food, the food exits the stomach through the pylorus, another muscle sphincter, into the small intestine.

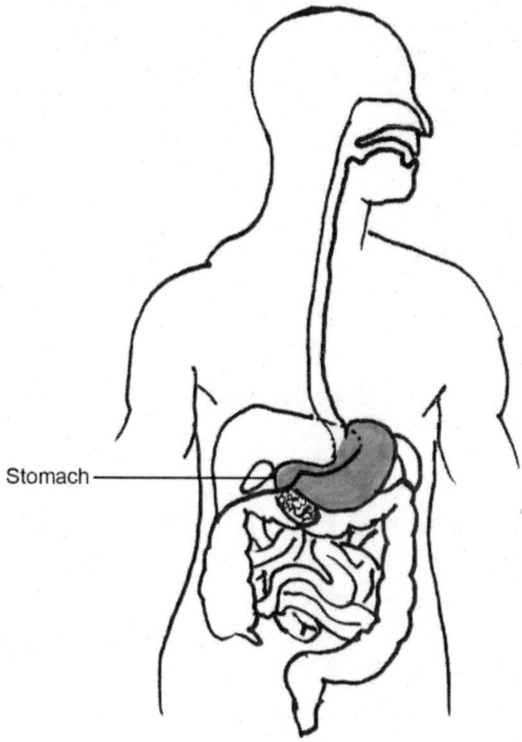

Figure 5.1. The schematic drawing shows where the stomach is situated with respect to the body and the rest of the digestive tract.

The stomach has a remarkable ability to expand in order to hold a lot of food. When it is empty, it has a volume of 50 milliliters. The stomach can expand to hold as much as 4 liters or about a gallon of food.

The wall of the stomach is made up of several layers. The inner most lining of the stomach is called the mucosa (Figure 5.3). When the stomach is empty, the mucosa forms small ridges or folds known as rugae. When the stomach is full, the mucosa is stretched out and the rugae flatten out to give the inside lining a smooth look. Underneath the mucosa is a layer called the submucosa, which separates the mucosa from the outer muscle layers of the stomach wall called the muscularis externa. The muscularis externa is made up of smooth muscle tissue which serves to move food along within the stomach.

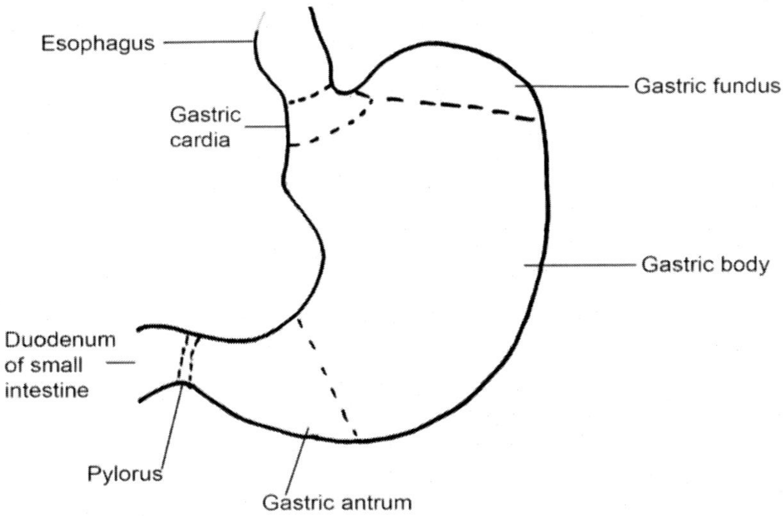

Figure 5.2. This schematic drawing demonstrates the location of the various regions of the stomach (cardia, fundus, body, antrum, and pylorus) and the relationship of the stomach to the esophagus and the duodenum of the small intestine.

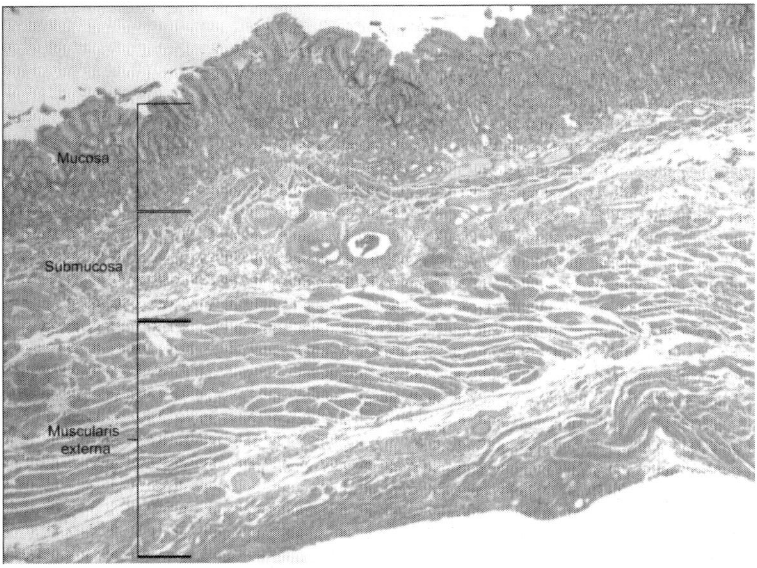

Figure 5.3. This is a picture of what the stomach wall looks like under the microscope, magnified 50 times. The mucosa, which is located on the inside aspect of the stomach, makes gastric juices which contain acids that contribute to food breakdown. The muscularis externa helps move the food along through the stomach toward the small intestine.

What Does the Stomach Do?

The stomach performs three major jobs for the digestive system. It stores the food that has been recently eaten. It begins to breakdown the food into a liquid mixture called chyme. After a few hours, the stomach then empties the liquid-like mixture into the next part of the digestive system, known as the small intestine. Lastly, the stomach serves as a mixer and masher. This mixing and mashing is performed by the muscularis externa muscles. Substances are produced by the stomach (like gastric juices) that chemically help to breakdown food. These juices also help to kill bacteria that may have been eaten with the food.

Where Do the Gastric Juices Come from?

The gastric juices are made by microscopic glands that are located in the mucosa of the stomach. The glands are made of up of several cells that are box-shaped and arranged in a ring. There are over 35 million glands in the stomach. There are different types of glands in the various regions of the stomach that are responsible for making different things. Some glands produce mucus, a fluid that protects the stomach lining from being hurt by the food and acids the stomach produces. Other glands produce hydrochloric acid and pepsin. Hydrochloric acid and pepsin are both important in breaking down proteins in the food.

What Causes Ulcers?

The strong acids that are made and released in the stomach can sometimes damage the inside lining of the stomach, also known as the mucosa, making a hole in the stomach mucosa known as an ulcer (Figure 5.4). Ulcers can cause stomach pain, vomiting, heartburn, and bleeding. It used to be believed that too much stomach acid was the cause of ulcers. The treatment was a bland diet and medicines that decreased the amount of stomach acid produced. We now know that there are bacteria in the stomach, called Helicobacter pylori, that play a significant role in the development of many ulcers. Medicines that kill these bacteria can help treat ulcers. Doctors can actually see ulcers by placing a long tube with a camera down the mouth and esophagus into the stomach.

The tube, also known as an endoscope, allows the doctor to look around at the mucosa of the stomach and search for ulcers and other lesions such as cancers.

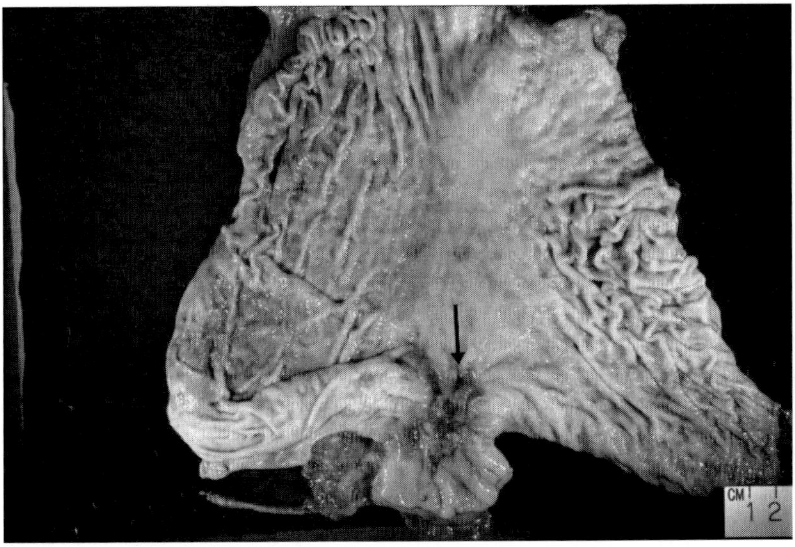

Figure 5.4. This is a picture showing what the inside of the stomach looks like. Note the area at the bottom (arrow). This represents an area where the mucosa of the stomach is damaged in a patient who has an ulcer.

What Causes a Stomachache?

A stomachache is a general term used for any kind of pain involving the stomach or intestines. There are a dozen of different things that can cause stomachaches. The most common culprits are infection (like you would get with food poisoning or stomach flu), ulcers, or heartburn (not really a stomach problem but rather more of an esophagus problem due to stomach acid moving back up into the esophagus.

What Causes Burping?

Burping or belching is caused by eating or drinking too fast. In the process, air is swallowed. The sound of burping is caused by the vibration of the cardiac sphincter (the muscular ring that closes between the esophagus and stomach) as gas is passed from the stomach to esophagus. Burping may also be caused by drinking carbonated beverages such as soft drinks, beer or champagne. Babies are particularly prone to swallowing air while feeding and need to be "burped" after eating or drinking to help get rid of the air. If not, the air in the baby's stomach may cause discomfort.

Stomach growling is caused when the stomach wall squeezes together to mix food and there is no food there. The gastric juices slosh around and cause the noise. The formal medical term for stomach growling is borborygmi.

Interesting Facts

1) Frogs can throw-up. When they do, they throw-up their stomach first. The stomach dangles out of their mouth. The frog uses it forearms to dig out the stomach contents and then the frog swallows the stomach back down. (www.scifacts.net/677/some-frogs-vomit-by-throwing - up...)
2) The starfish is able to turn its stomach inside out and extend it partially outside the body. (www.youtube.com/watch??v-fxzUtnKTN QE)
3) Crayfish produce stones made up of calcium in their stomach. When the animal sheds its outside shell, the stones are reabsorbed by the stomach and the calcium is used to form a new shell. (www.britannica.com/EBchecked/topic/567085)
4) Camels regurgitate the contents of their stomach and spit it at their enemies. (www.animals.sandiegozoo.org/animals/camel)
5) Ants have two stomachs. One holds food for itself and a second stomach holds food to be shared with other ants. (www.lingolex. com/ants.htm)
6) Sometimes, dinosaurs used to deliberately swallowed stones. These stones or gastroliths, some as large as footballs, would sit in the stomach and assist in crushing and churning plant food in order to help break the food into smaller pieces. (www.rareresources.com /gastroliths.htm)
7) In 1927, a 42-year-old woman complained of mild abdominal pain and had over 2533 objects removed from her stomach including 947 pins. (listafterlist/top-50-weirdest-guinness-world-records)
8) Can you train your stomach? In 2000, Sonya Thomas ate 26 grilled cheese sandwiches in 10 minutes. She also holds several other world

eating records including 46 dozen oysters in 10 minutes, 11 pounds of cheesecake in 9 minutes, 37 hotdogs with buns in 12 minutes and 56 hamburgers in 8 minutes. She said she had a naturally big stomach capacity and to train, she drinks large amounts of water to expand her stomach capacity. (en.wikipedia.org/wiki/Sonya-Thomas)

9) The loudest burp was made by a man at over 100 decimals. A large orchestra produces noise at 98 decimals.
(www.youtube.com/watch?v=MLGDSpj7EHI)

Chapter 6

Small Intestine

Where Is the Small Intestine?

The small intestine is located below the stomach in the abdominal cavity (Figure 6.1). The contents of the stomach are emptied directly into the small intestine. The small intestine is the longest segment of the digestive system (4-5 times longer than the large intestine). It is compressed into many folds in order to take up the least amount of space in the abdominal cavity. The reason why it is called the "small" intestine is that the diameter of the small intestine is much smaller than that of the large intestine.

What Does the Small Intestine Look Like?

The small intestine looks like a long tube and it is divided into three parts: 1) the duodenum, which is the shortest segment and the part that receives the chyme (semifluid material produced by the stomach that represents partially digested food) from the stomach, 2) the jejunum, which is the middle segment and 3) the ileum, which is the last and longest segment.

Like the stomach, the small intestine consists of an inner mucosal layer made up of glands, a submucosal area, and outer muscle layers called muscularis externa (Figure 6.2). The inside lining of the small intestine is different than the stomach because it is made of numerous folds called pilcae circulares or circular folds. Each fold or plica is covered by numerous tiny projections called villi (Figure 6.3).

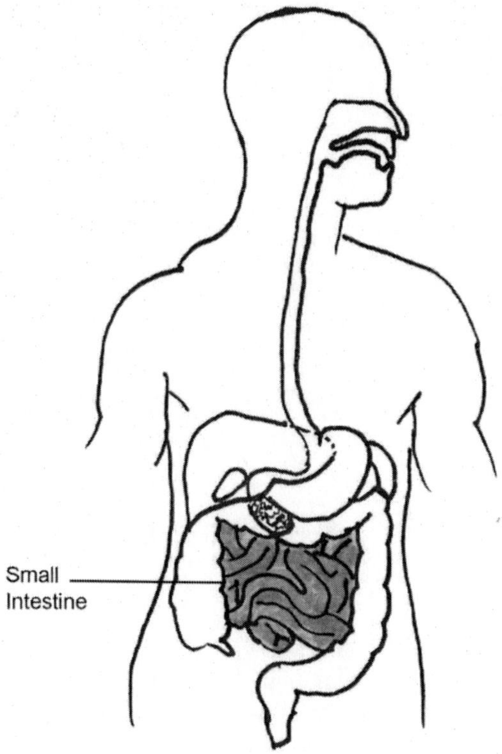

Figure 6.1. This schematic drawing highlights the location of the small intestine with respect to the rest of the body and the rest of the digestive system.

Why All the Folds and Villi?

One of the major functions of the small intestine is to absorb nutrients. The folds and villi increase the surface area of the small intestine in order to help increase its absorptive capacity. Almost 90% of the absorption of nutrients occurs in the small intestine. The food is moved along the small intestine by peristalsis, just like the esophagus. As the food is moved through the small intestine, the cells on the surface of the villi take in nutrients. The nutrients will then be later distributed throughout the body. Most of the absorption or taking in of nutrients occurs in the jejunum. Many of the vitamins, minerals and nutrients discussed in the last chapter are absorbed at different locations along the small intestine. For example, substances like iron are absorbed in the duodenum. Vitamin B12 is absorbed in the ileum.

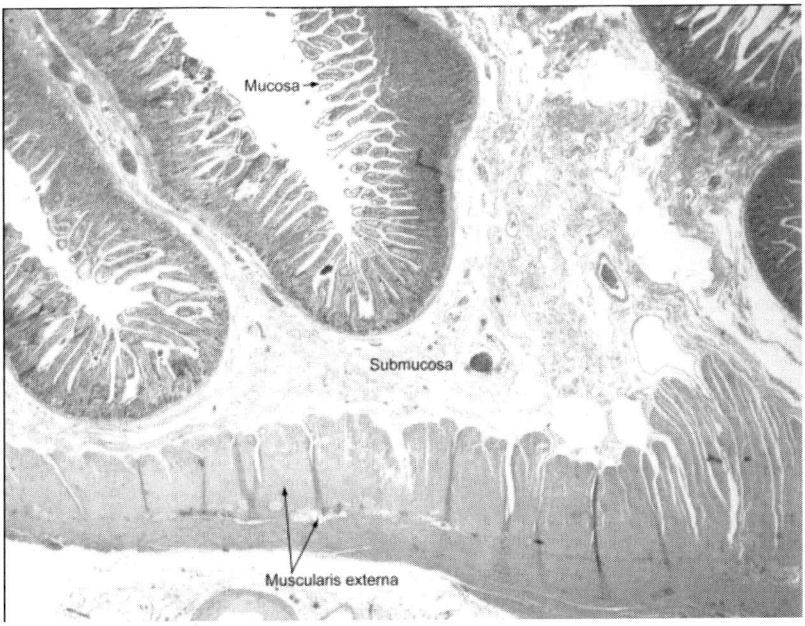

Figure 6.2. This is a picture of what the duodenum of the small intestine looks like under the microscope, magnified 20 times. This image shows all three layers of the duodenal wall. The inside mucosa made up of papillary structures, the submucosa, and the muscularis externa muscle layers.

In addition to absorption, the glands and the mucosa of the small intestine make other substances. They produce mucus to help protect the small intestine from the stomach acids that enter with the food. Additionally, substances made by the small intestine, as well as substances which enter the small intestine from the pancreas and liver (see chapters 7 and 8), help breakdown food into even smaller particles that can be absorbed. Proteins, fats, and sugars are all broken down into small particles in the small intestine.

What Is Lactose and Lactose Intolerance?

Lactose is a type of sugar that is found in certain foods like milk, yogurt, and ice cream. Normally, the lactose sugar is broken down into smaller particles by a protein called lactase, which is made in the small intestine. In people who do not make a lot of lactase, the lactose sugar in milk or yogurt cannot be properly broken down. When there is too much undigested lactose around in the gastrointestinal system, the person might experience stomachaches, diarrhea, or increased gas. These symptoms are related to the fact that the undigested lactose is broken down by bacteria in the gastrointestinal tract, resulting in the formation of gas. This condition of not being able to properly digest lactose sugar is called lactose intolerance. There is no cure for lactose intolerance. Usually, people have to avoid drinking or eating food with lots of lactose and consume lactose-free food products. There are also oral pills that contain lactase which can serve as a means of replacing the lactase that is missing in the small intestine.

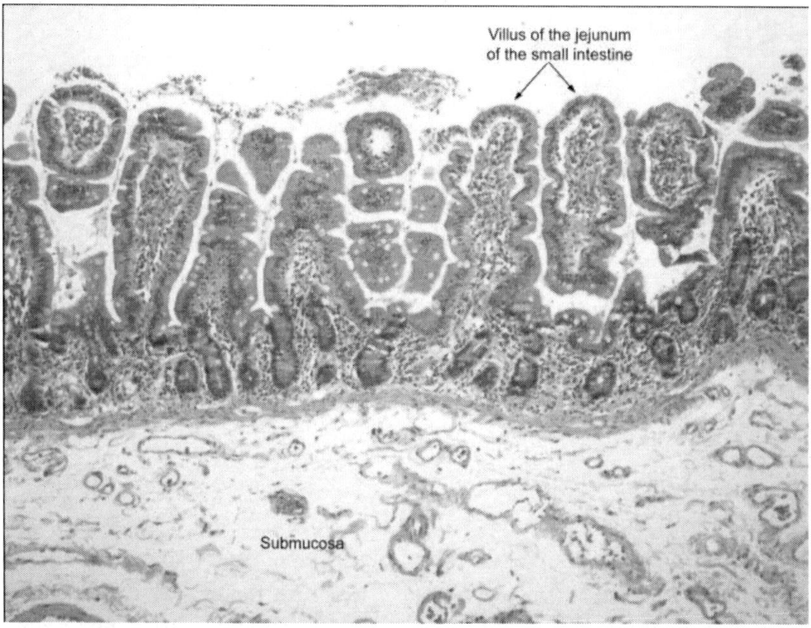

Figure 6.3. This is a picture of what the jejunum of the small intestine looks like under the microscope, magnified 100 times. Each of the papillary structures in the mucosa represent a villus. Each villus is lined by cells which function to absorb nutrients. The villus architecture increases the surface area available for nutrient absorption.

Interesting Facts

1) The entire small intestine stretches out to a length of approximately 22 feet and 8 inches in an adult. (en.wikipedia.org/wiki/Small-intestine)
2) Food passes through the stomach and small intestine in about 6-8 hours, (www.mayoclinic.org/digestive-system/expert-answers/FAQ...)

3) Sharks and sturgeons do not have a small intestine. (en.wikipedia.org/wiki/Small-intestine)
4) Improperly cooked beef or pork may contain tapeworms. Tapeworms are organisms, known as parasites, which can attach to the wall of the small intestine and live off the partially digested food in the small intestine. Tapeworms can grow up to 15-30 feet in length! (en.wikipedia.org/wiki/Tapeworm-infection)

Chapter 7

Liver and Gallbladder

Where Are the Liver and Gallbladder Located?

Both the liver and the gallbladder are located just below the diaphragm and can be found on the right side of the body (Figure 7.1). More specifically, the liver sits to the right of both the stomach and the pancreas (which we will discuss in the next chapter). The gallbladder is connected to the liver by a tube called the hepatic duct. Hepatic is a word that simply describes something associated with the liver. The gallbladder can be found just below the liver (Figure 7.2).

What Do the Liver and Gallbladder Look Like?

The liver is the largest organ in the abdominal cavity. It is made up of 4 distinct parts called lobes. The two largest are called the right and left lobe. The two smaller lobes are called the quadrate and caudate lobes. The liver tissue has a dark brown appearance (Figure 7.3).The gallbladder is a small, green-colored organ that looks like a pouch.

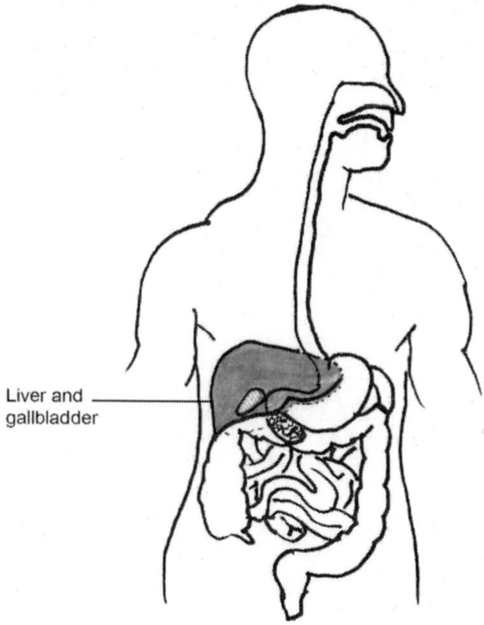

Figure 7.1. This schematic drawing shows the relationship of the liver and gallbladder to the rest of the body and digestive system.

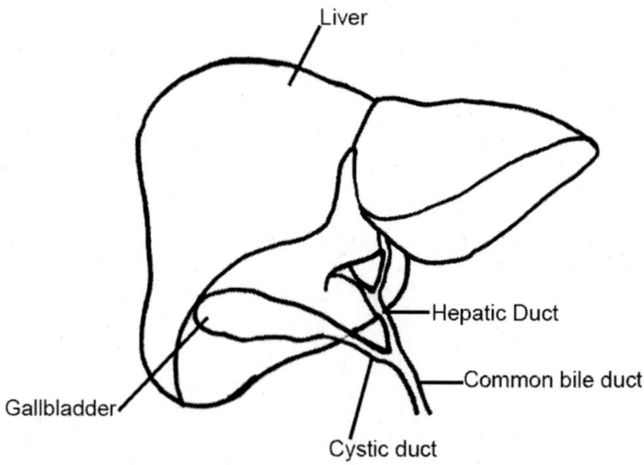

Figure 7.2. This schematic drawing illustrates the relationship of the gallbladder to the liver. The gallbladder's contents are emptied into the gastrointestinal tract via the cystic duct. The contents of the cystic duct are combined with the contents of the hepatic duct (coming from the liver) in the common bile duct, which empties into the small intestine (duodenum).

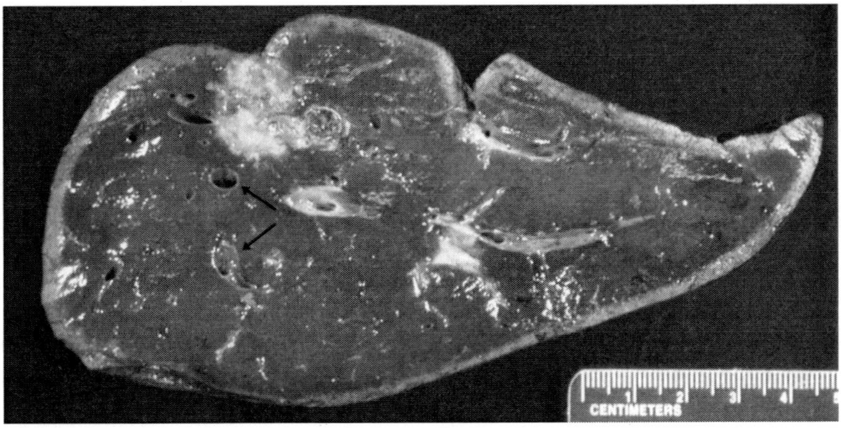

Figure 7.3. This is a picture showing part of a normal liver. It has a dark brown appearance. The holes (arrows) represent blood vessels that normally make up part of the liver tissue.

What Does the Liver Do?

The liver has important duties that keep you alive and healthy. It functions as the primary organ in metabolism. Metabolism is a broad term that describes all the ways in which the body creates or uses energy and how the body can maintain its functions. Blood from the body's circulatory system (a system which consists of blood vessels (arteries and veins) and blood cells) pools in the liver through the venous blood system. Specifically, the blood enters the liver through the hepatic portal vein. Notice the word hepatic is once again used to describe the vein's association with the liver! The blood carries many different molecules like vitamins, minerals, sugars, amino acids and fats that have been absorbed by the digestive system. The liver is able to modify and coordinate how the body will use these molecules. Additionally, the liver is important in producing energy from proteins and fats, if you have not eaten for a while.

What Is Cirrhosis of the Liver?

Cirrhosis is characterized by scarring of the liver due to disease or injury. It typically occurs over lengthy period of time and results in the liver being unable to do its job effectively. There are many causes of cirrhosis. One of the common causes of cirrhosis is hepatitis, which is often caused by an infection of the liver due to a virus. Breaking down the word hepatitis, we can understand what it means. If you remember, "hepatic" means it is associated with the liver and "-itis" simply means inflammation. This means that hepatitis is inflammation of the liver. This inflammation is often caused by the body's attempt to fight off the viral infection.

The liver also plays an important role in eliminating toxins (poisonous chemicals) from the body. For example, the liver is responsible for getting rid of alcohol when someone drinks an alcoholic beverage. If someone drinks a lot of alcohol for a long time, he or she can also develop cirrhosis amongst other health problems (Figure 7.4).

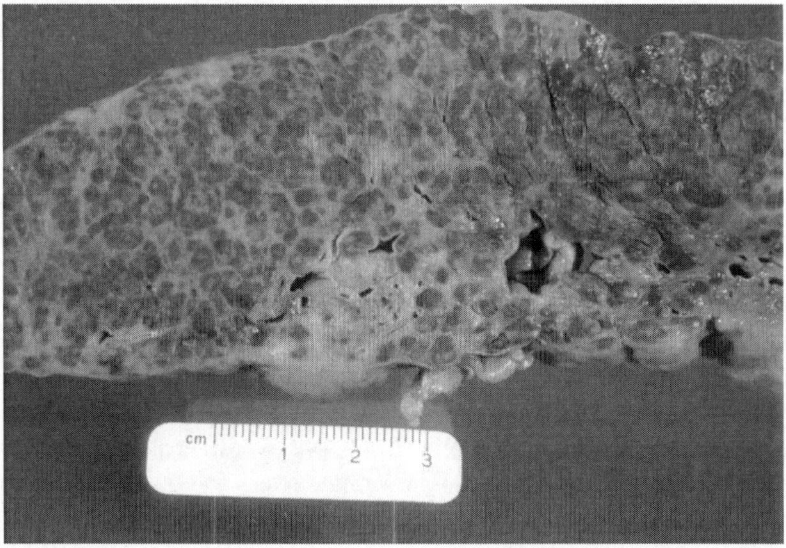

Figure 7.4. This picture is taken of a liver from a patient who abused alcohol. There is increased scar tissue (lighter appearing) in the liver (cirrhosis). The darker nodules are green colored (normal liver is dark brown) and is due to abnormally increased bile in the tissue.

What Is the Bile?

The liver is also responsible for making bile. Bile is important in the digestion process of fats, which are also called lipids. Bile leaves the liver through a short system of ducts or tubes, which dump the bile into the gallbladder (Figure 7.5). The gallbladder is a storage organ for the bile produced by the liver, until you eat fatty foods. When you eat food-containing fat, the bile leaves the gallbladder through the cystic duct and common bile duct. The bile ultimately finds its way into the duodenum of the small intestine, where it can help break up or emulsify the fat that you just ate. This helps the small intestine absorb fats by breaking them into smaller pieces and preventing them from joining together.

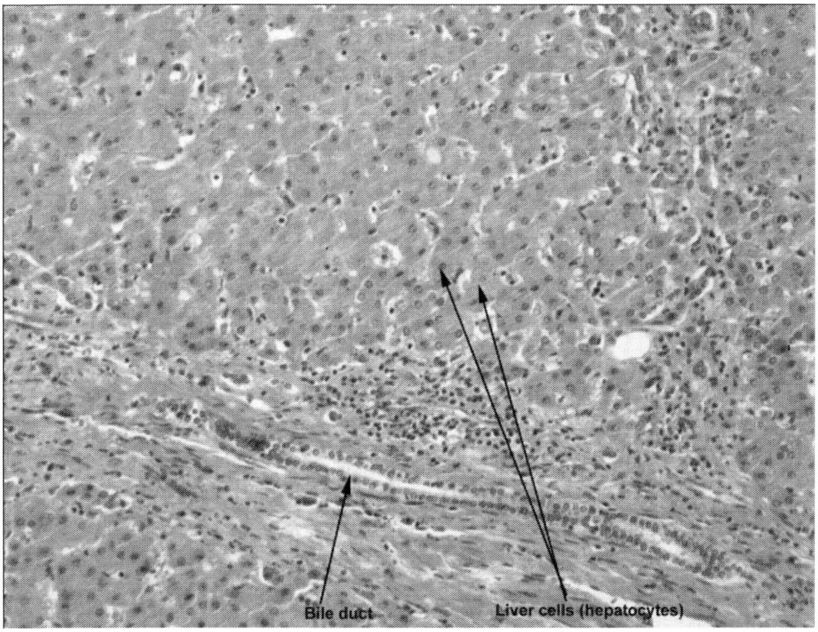

Figure 7.5. This is a picture of what the liver looks like under the microscope, magnified 200 times. Each of the bigger round cells is a liver cell, also known as a hepatocyte. The elongated, tubular structure represents a bile duct. The bile duct system transports the substances made by the liver cells, including bile, outside the liver to the rest of the body.

What Are Gallstones?

Gallstones are like small rocks that sometimes develop in the gallbladder (Figure 7.6). There are two major types of gallstones. Specifically, there are cholesterol-based gallstones and there are bilirubin-based gallstones.

Cholesterol is necessary for the transport of different types of fat throughout the body and is a type of fat itself. Bilirubin is a waste product made when old red blood cells are broken down. Gallstones occur when these compounds become tightly bound together to form something that looks and feels like a pebble. They can vary to some degree in size, but are often about the size of a marble. They can be found in the gallbladder itself or in the duct system leading from the gallbladder to the duodenum. They are often quite painful, especially when they get stuck in the duct system, and can sometimes significantly disrupt the flow of bile into the duodenum. This disruption can thus make it difficult for the body to properly digest fats. The blockage can also lead to a life-threatening condition called pancreatitis which will be discussed in the next chapter.

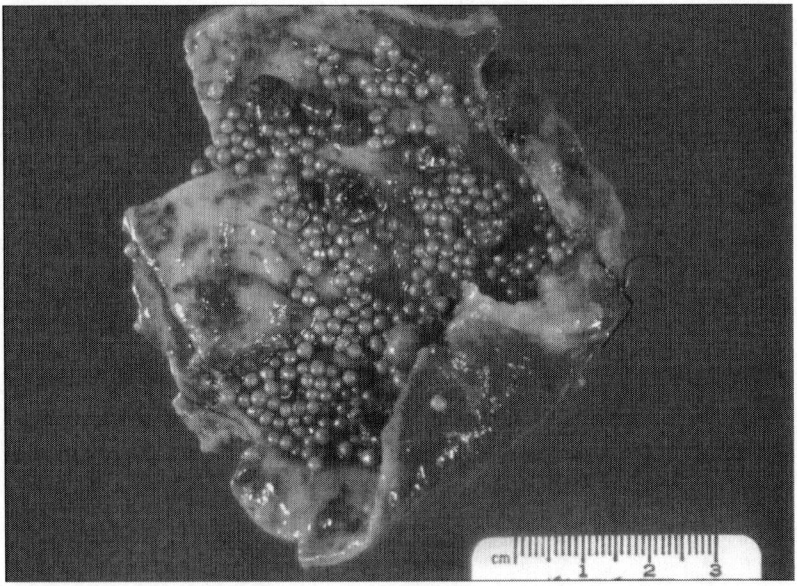

Figure 7.6. This picture shows the inside of a gallbladder filled with a lot of gallstones. The stones are of two types: the small round white ones that look like pearls and the larger dark black ones.

Interesting Facts

1) In Greek mythology, Prometheus was punished by the gods for giving fire to humans by being chained to a rock where a vulture would peck at and eat his liver. The liver would grow back overnight and the

vulture would return the next day to peck and eat at it again. (en.wikipedia.org/wiki/Prometheus)

2) The ancient Greeks thought the liver was the place where emotions like wrath, jealousy and greed originated. (www.factslegends.org/25-facts-human-liver)
3) The liver is the second largest organ in the body, after the skin. (www.liverfact.com)
4) A number of animals do not have gallbladders including rats, deer, and horses. (en.wikipedia.org/wiki/Gallbladder)
5) In 1952, a 13.84 pound gallstone was removed from an 80-year-old woman in London. (gallbladderhelp.com/worlds-largest-gallstone)

Chapter 8

Pancreas

Where Is the Pancreas Located?

The pancreas is an important hormone producing organ located in the upper left portion of the abdomen. More specifically, the pancreas is actually located behind the stomach and next the first part of the small intestine (duodenum) (Figures 8.1 and 8.2).

What Does the Pancreas Look Like?

The pancreas has a round head-like component and a thinner tail. The head is located more midline to the tail of the pancreas. It is connected to the small intestine by a duct system. (Figure 8.3). Specifically, the pancreas has a pancreatic duct that connects with the common bile duct to form the ampulla of Vater. This is the structure that merges with the wall of the small intestine. It is through these ducts that some of the digestive chemicals created in the pancreas can enter the small intestine and help to breakdown the food located there.

What Does the Pancreas Do?

The pancreas produces some chemicals, such as trypsin or proteases that help breakdown the food in the small intestine. These chemicals travel through the pancreatic duct system and the ampulla of vater to enter the small intestine. Additionally, the pancreas plays a vital role in producing hormones that the body needs to help regulate its processes. Specifically, the pancreas produces two hormones that are used to maintain the amount of glucose in your blood. The circulatory system is the body's means for transporting nutrients to most of the tissues in the body. Since glucose, a simple sugar, is the main fuel source for the body's cells, it is important that you are able to regulate the amount of glucose being transported around the body and the amount of glucose being taken in by various cells throughout the body!

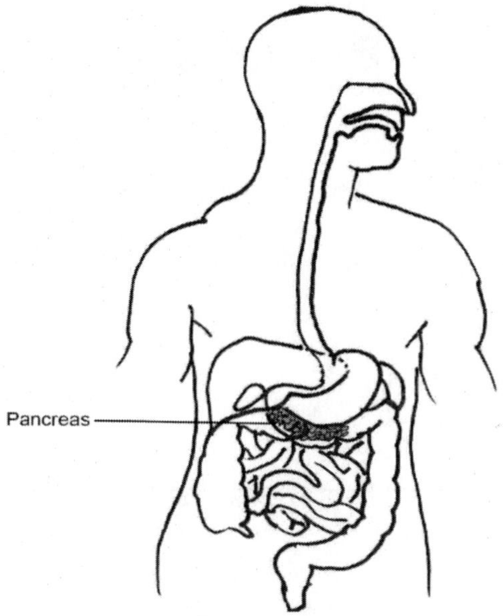

Figure 8.1. This schematic illustration shows where the pancreas is located in relation to the rest of the body and the digestive tract.

The first of these important hormones is insulin. Insulin is produced by the beta cells, located in the islets of Langerhans, of the pancreas (Figure 8.4). Insulin's main function is to reduce the amount of glucose in the blood by signaling cells to take in glucose from the blood. Insulin is released by the pancreas after you eat a meal to deal with the glucose that will be made from the breakdown of the food that you just ate. The other hormone is called glucagon. It is produced in the alpha cells of the pancreas, also located in the islet of Langerhans. It has the opposite effect of insulin and signals the release of glucose from cells into the blood. So overall, insulin reduces the amount of glucose in the blood and glucagon increases the amount of glucose in the blood.

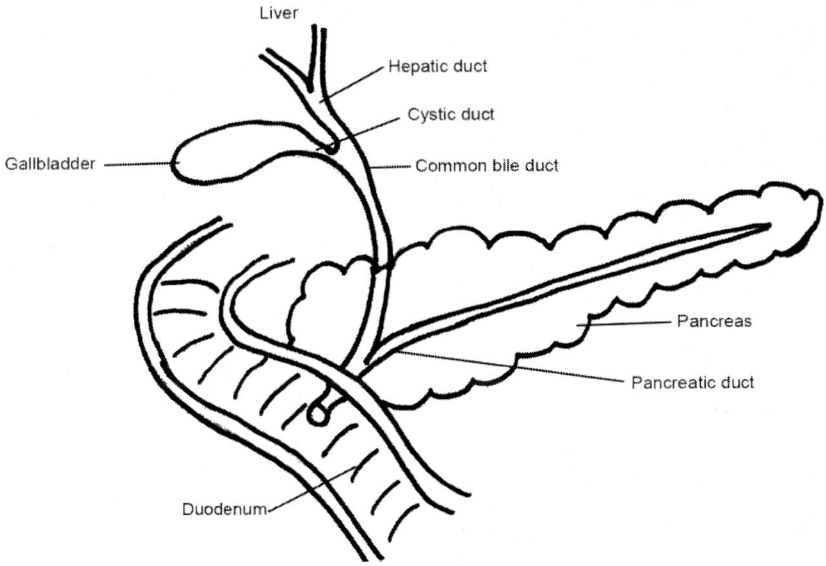

Figure 8.2. This drawing highlights the relationship of the pancreas to the liver, gallbladder and duodenum. The contents of the pancreas that aid in digestion are moved from the pancreatic duct toward the duodenum.

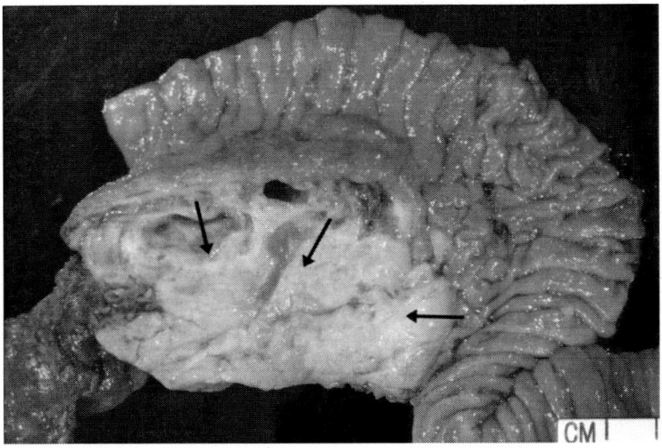

Figure 8.3. This picture shows what the inside of a normal small intestine (top) looks like and what the relationship of the normal pancreas (bottom, arrows) to the small intestine is.

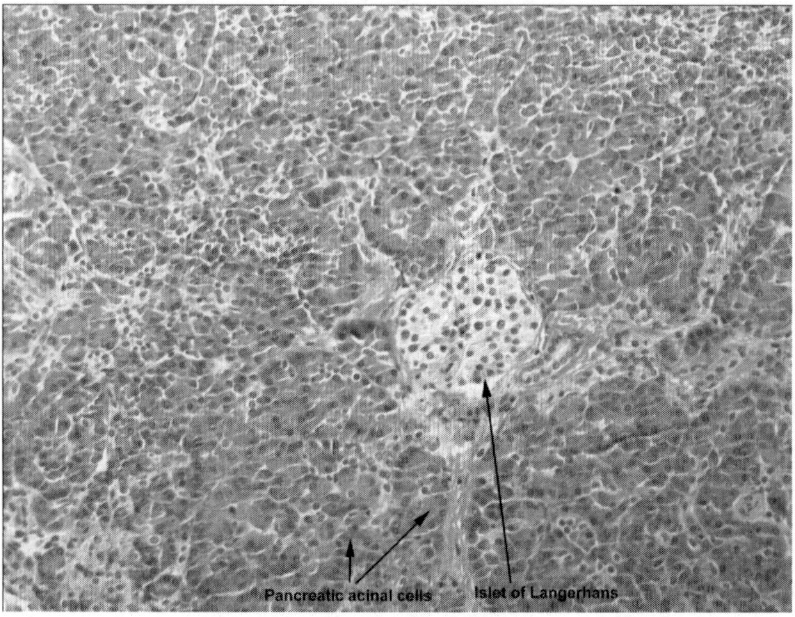

Figure 8.4. This is a picture of the pancreas looks like under the microscope, magnified 200 times. The cells in the islet of Langerhans are responsible for making hormones like insulin and glucagon. Most of the cells seen here represent acinar cells which produce substances that are released into the small intestine and help in food digestion.

What Is Diabetes?

Diabetes is a disease in which the pancreas is unable to properly produce insulin. There are two main types of diabetes, type 1 and type 2.

Type 1 diabetes is caused by the auto-immune destruction of the beta cells of the pancreas. As we mentioned before, these beta cells are responsible for the production of insulin, so their destruction means that insulin will not be produced. "Auto-immune" means that the person's own immune or defense system is attacking the beta cells. The reason why these people's own defense system would attack their own cells is still not completely understood and is an area of current research.

Type 2 diabetes is an acquired form of the disease. It is often caused by many different factors that together cause the development of diabetes. These factors include, lack exercise, poor diet, and some genetic elements. When someone has high blood glucose levels, the body tries to adjust to this situation. Normally, the body is able to make this adjustment by manufacturing more insulin in order to reduce the amount of glucose in the blood. However, if someone has high levels of glucose for long periods of time (months or years), her or his body may eventually become unable to keep up with the increased need for insulin. In these individuals, their insulin producing beta cells are still present and they are not able to produce the amount of insulin needed to reduce the high glucose levels.

Diabetes has potentially harmful effects all over the body, including effects on the eyes, nerves, blood vessels, heart, and kidney. If a diabetic person is not able to manage their disease, then they can have serious health problems. Diabetes is not a disease that can be cured, but it can be managed. Individuals can live their entire lives with diabetes, so long as they manage their illness properly. Maintaining a healthy diet and exercise are important tools in dealing with diabetes. People with Type 1 diabetes typically have to inject themselves with insulin multiple times a day or use an insulin pump in order to supply their body with enough insulin to manage their blood glucose levels. People with Type 2 diabetes sometimes will use similar treatment methods, but more often they will be given medications that will help them maintain appropriate blood glucose levels with the insulin their own beta cells are producing. In either case, it is important for diabetics to monitor their blood glucose levels throughout the day in order to ensure that their levels are within a healthy range. Diabetics also must keep track of the amount of sugar in the food that they eat in order to make sure that they maintain a healthy blood glucose level.

What Is Pancreatitis?

Pancreatitis is another word with"-itis" in it, which means that it describes inflammation. In this case, it refers to inflammation of the pancreas! There are several potential root causes of pancreatitis including, gallstones, alcoholism, or viral infection. These factors can result in the pancreas' inability to properly secrete its digestive enzymes into the small intestine. As a result, these enzymes can damage the pancreas itself, which can result in serious illness and sometimes death, if not treated early enough (Figure 8.5). Depending upon the cause, pancreatitis can develop rapidly or over the course of many months to years.

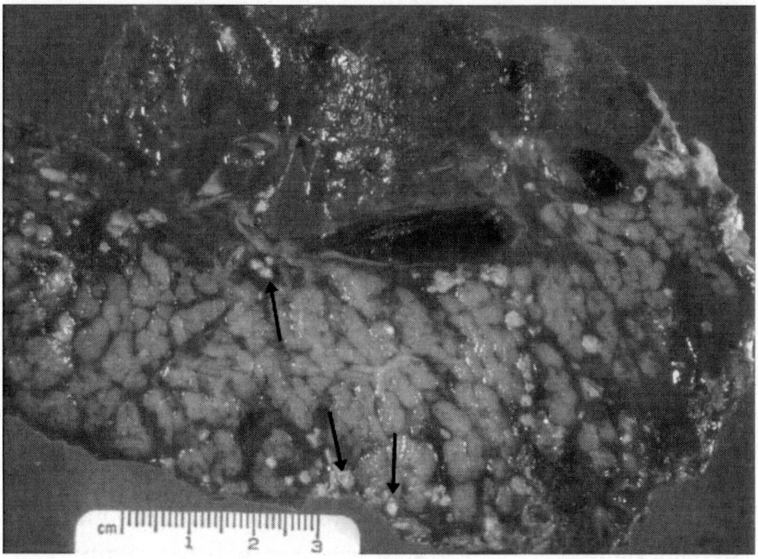

Figure 8.5. This pancreas is inflamed and has adarker, typically red appearance in areas. Small areas of dead pancreatic tissue are also present (arrows). This pancreas was from a patient who had pancreatitis.

Interesting Facts

1) Diabetes is the most common endocrine disorder in the United States, affecting 9.3% of the population or 29.1 million people. (www. cdc.gov/diabetes statisticsreports.html)
2) The earliest descriptions of diabetes date back to antiquity. An Egyptian manuscript from c.1500 BCE mentions increased urine production with diabetes. In India at about the same time, the disease was known as “honey urine” because it was observed that the urine would attract ants. (en.wikipedia.org/wiki/History-of-diabetes)
3) Rabbits do not have a well-defined pancreas. Instead, they have pancreatic tissue diffusely distributed in the abdomen. www.news-medical.net/health/What-Does-The -Pancreas-Do)

Chapter 9

Large Intestine and Rectum

Where Is the Large Intestine?

The final stretch of the digestive system slide involves the large intestine and rectum (Figure 9.1). The large intestine, also known as colon, is located on both sides of the abdominal cavity. The large intestine is shorter than the small intestine, but has a larger diameter.

What Does the Large Intestine Look Like?

The large intestine is tube, which like the small intestine is divided into several parts. The ileum portion of the small intestine empties its contents into the right side of the large intestine, which includes the cecum and the ascending colon. The cecum is a short segment of the large intestine that is next to the ileum and is located in the lower right corner of the abdomen. The contents of the cecum proceed into the ascending colon. The ascending colon is located behind the rest of the other abdominal organs in an area known as the retroperitoneum. The first part of the ascending colon, as its name suggests, starts low down in the abdomen and extends upward toward the liver.

The next segment of the large intestine is called the transverse colon. It is situated in a horizontal fashion, running from the liver on the right side toward the spleen on the left.

From the transverse colon segment, the large intestine takes a downward turn into the descending colon. The descending colon travels down the left side of the abdomen and connects with the sigmoid colon segment. The name sigmoid means S-shaped, since this portion of the colon is shaped in roughly an "S" configuration.

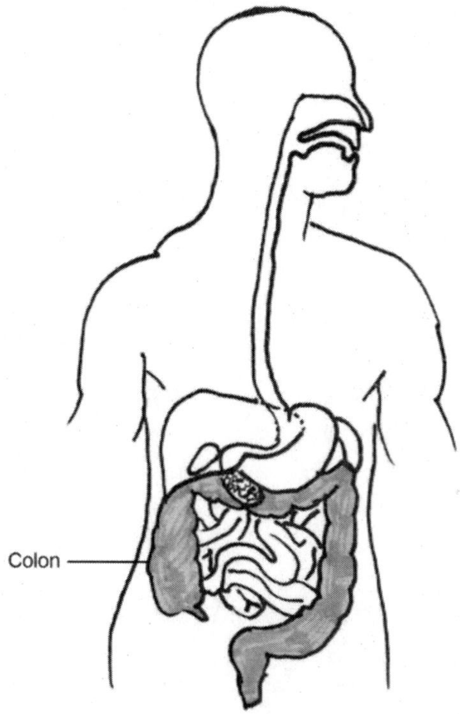

Figure 9.1. This schematic drawing shows the relationship of the large intestine (colon) to the rest of the body and digestive system.

The final segment of the large intestine is the rectum. The rectum is a portion which is about one foot in length and it holds the feces or stool before exits the rectum through the anus. The anus is the external opening of the rectum.

Similar to the small intestine and stomach, the wall of the large intestine consists of three layers: an inner mucosal layer containing tubular structures called glands, a submucosal layer, and an outer smooth muscle layer (Figures 9.2 and 9.3). Unlike the small intestine, the mucosa is much flatter and is not comprised of villous structures.

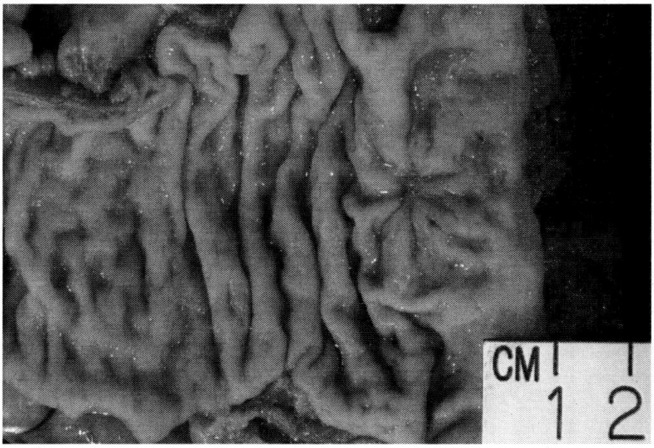

Figure 9.2. This picture shows what the inside of a normal colon and rectum look like. Note the folds in the mucosal tissue.

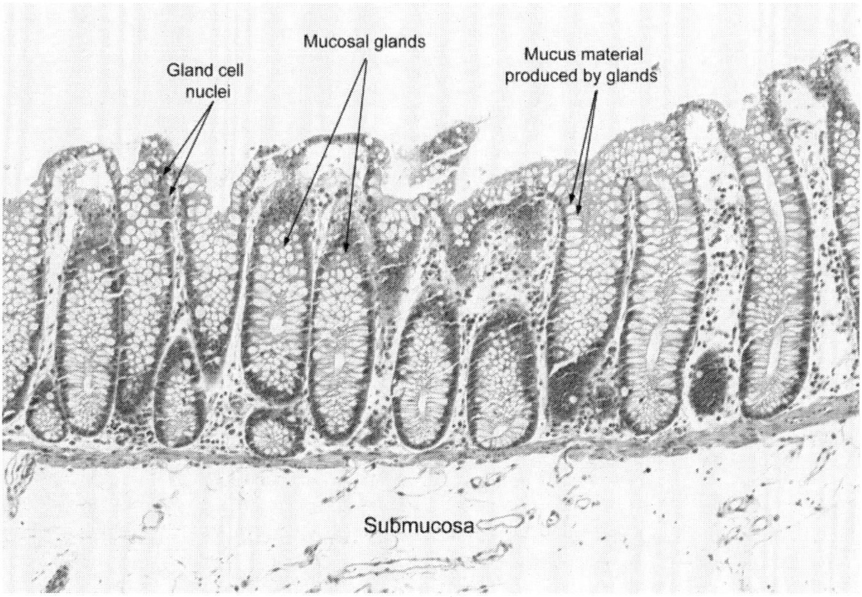

Figure 9.3. This is a picture of the colon looks like under the microscope, magnified 200 times. Each of the elongated tubular structures represents a mucosal gland. The darker structures are the nuclei that belong to the cells lining the tubules. The white areas next to each nucleus in the cells is mucous material. This mucus gets released from the cells and is helpful in moving the chyme along.

What Does the Large Intestine Do?

The large intestine serves several functions. It is responsible for removing the remaining water from the mostly digested food. What remains in the large intestine after the remaining water is removed is cellulose (a plant sugar that makes up the cell wall), some undigested materials and other digested materials. The colon is also rich in bacteria, which live on the remaining undigested materials. Some of the bacteria can produce nutrients, like vitamin K, that can be absorbed by the colon and used by the body. The glands in the colon's mucosa produce a mucous substance that helps the chyme move along (Figure 9.3). The indigestible material and bacteria eventually move into the rectum and exit through the anus as stool.

Where Is the Appendix and What Does It Do?

The appendix is a small worm-like tube that is attached to the cecum of the large intestine (Figure 9.4). In humans, it serves little function but in some animals it is involved in digesting cellulose material. In some humans, the appendix may become infected or inflamed, causing something called appendicitis (Figure 9.5). This usually causes a lot pain. The appendix, in this case, needs to be removed with surgery. The surgical removal of the appendix is called an appendectomy.

What Causes Diarrhea?

Most people have experienced the unpleasantness of diarrhea. Diarrhea is a condition in which one has bowel movements more frequently; the stool or feces is sometimes very watery and not well-formed. Remember, both the large and small intestine are responsible for reabsorbing water. When something happens that interferes with the reabsorption, the water stays in the stool, making it runny. Because a lot of water can be lost if there are a lot of bowel movements, it can be potentially life-threatening, especially in babies. There are a number of things that can cause diarrhea by interfering with waster reabsorption. The most common causes are infections by either viruses or bacteria.

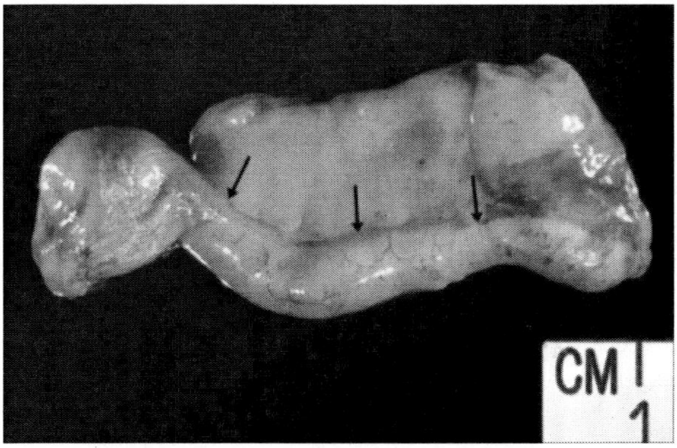

Figure 9.4. The worm-like structure seen here (arrows) is what the outside of a normal appendix looks like. The tissue at the top is fat or adipose tissue, which is normally attached to the outside surface of the appendix.

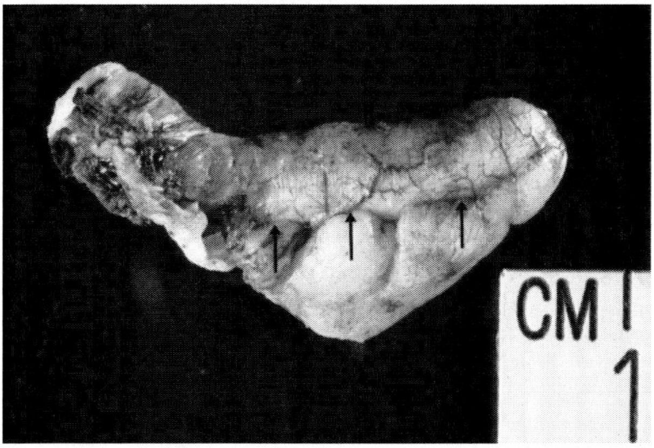

Figure 9.5. The appendix illustrated here (arrows) looks enlarged (swollen) and is abnormally colored. This patient had inflammation of the appendix, known as appendicitis.

What Cases Farting and Why Does It Sometimes Smell So Bad?

Farting is the passing of gas (flatulence) through the anus. Where does the gas come from? Most of the gas comes from bacteria in the large intestine, as they try and digest cellulose and the other substances that our bodies cannot normally digest. Certain foods such as beans, mushrooms, cabbage, and onions have material that is not easily digested by humans, and therefore, these foods can serve as sources of food for these gas producing bacteria.

Why the smell? Not all flatulence has an odor associated with it. The flatulence is mostly made up of five gases: nitrogen, oxygen, carbon dioxide, hydrogen and methane. The hydrogen can sometimes combine with sulfur from certain foods such as eggs or cauliflower to make a substance known as hydrogen sulfide. Hydrogen sulfide is responsible for the terrible odor that is associated with flatulence. The noise sometimes associated with flatulence is created by gas passing through the tightly constricted anal sphincter.

What Is Constipation?

Constipation is a condition in which a person has hard stool that is difficult to move. It may cause discomfort and pain. The condition is more common in children and older adults and can be caused by dehydration (not enough water in the body) or low fiber diet. Fiber refers to plant cell wall material like cellulose, which usually helps move food through the digestive tract. Treatment of constipation often involves drinking more fluids, especially water, and increasing the amount of fiber in the diet. One can increase dietary fiber by eating more vegetables, like broccoli. Some people may require additional treatment and the use of laxatives which are foods or drugs that help move the contents of the intestine through more easily.

What Cases Colon Cancer?

Cancer is an abnormal growth that occurs anywhere in the body. In the digestive system, the most common place for a cancer or tumor to arise is in the colon (Figures 9.6 and 9.7). Colon cancer tends to present in older adults. People with colon cancer may develop diarrhea, blood in their stool, weight loss, or pain. In people who are suspected of having colon cancer, a tube called a colonoscope can be inserted into the rectum, which allows the doctor to look inside the colon. If colon cancer is found, it can be treated by cutting out the tumor during surgery, by giving certain drugs to kill the tumor (called chemotherapy) or giving radiation treatments (use of high energy wave particles to destroy cancer cells). There is no single cause for colon cancer. Some cases appear to be influenced greatly by the results of an individual's genetics and there is also an increased risk of developing colon cancer if an individual has had a previous inflammatory lesion.

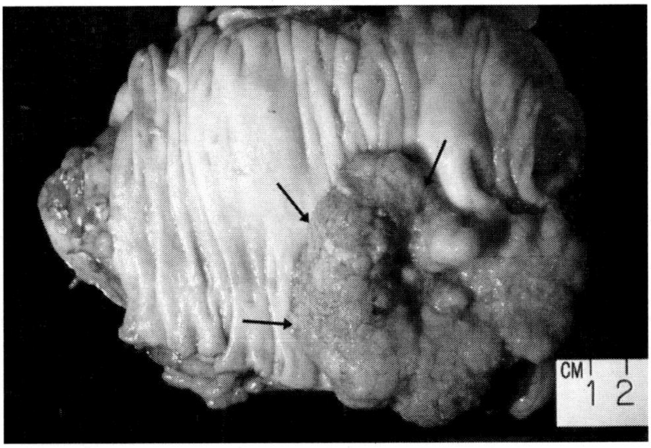

Figure 9.6. This picture shows a large growth in the colon which represented a colon cancer (arrows).

Inflammation is the basic way the body reacts to infection, irritation, or injury. There are three main types of inflammatory diseases that can affect the colon and each of these are associated with colon cancer. One of these diseases is known as Crohn's disease. Crohn's disease causes inflammation involving the entire thickness of the intestinal wall, including the mucosa, submucosa, and smooth muscle layers. Crohn's disease can affect any part of the tubular intestinal tract (esophagus, stomach, small intestine, large intestine, or appendix). A second inflammatory disease that involves the large intestine and rectum alone is ulcerative colitis. This disease only involves the mucosal layer of the intestine. The third main type of inflammatory disease that can affect the colon is caused by an infection. A variety of different organisms can cause an infection in the colon including different types of bacteria, viruses, fungi, and parasites.

What Are Hemorrhoids?

Hemorrhoids describe swelling and inflammation of certain types of blood vessels called veins in the rectum or anus. The development of hemorrhoids is related to straining too much while trying to pass stool or putting too much pressure on the veins due to poor muscle tone. Sitting for long periods is an example of this. Hemorrhoids can be painful, itchy and could cause bleeding.

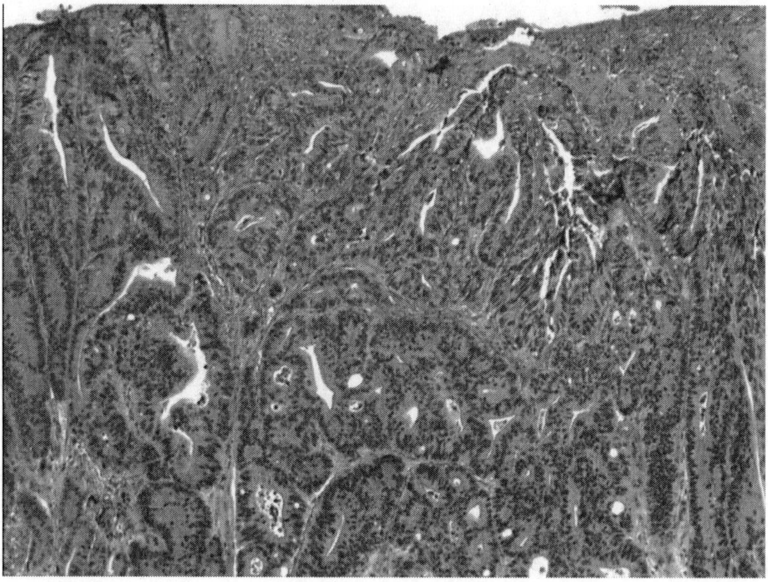

Figure 9.7. This is a picture of what colon cancer looks like under the microscope, magnified 100 times. Compare this picture to Figure 9.3. The cancer shows a disordered, abnormal arrangement of glands.

You have now completed your trip down the digestive system slide. The last chapter in this book will review nutrients. Nutrients are those substances in food that supply energy and the raw materials your body needs to survive, grow, and repair itself.

Interesting Facts

1) The average adult produces about 500-1500 ml of fart gas per day. (www.vox.com/2014/12/4/733411/fart-flatulence)
2) One animal's feces is another's food. Termites eat the feces of other termites (en.wikipedia.org/wiki/Coprophagia).
3) The Sonaran coral snake, when it feels threatened, sends out a clear warning sound - it farts. (www.anapsid.org/snakefart.html)

4) Ancient Egyptians noted the appendix during their mummification process. They referred to the appendix as the "worm of the bowel". (http://laparoscopyappendectomy.html)
5) Constipated bears during hibernation? Bears form an intestinal blockage called a tappen during hibernation. This prevents the hibernating bear from pooping in its den. (en.wikipedia.org/wiki/Tappen)
6) During the 1980 World Series, George Brett left game 2 in the 6th inning due to hemorrhoid pain. He had minor surgery the next day and in game 3 hit a home run. His team won game 3 by a score of 4 - 3 in 10 innings. (en.wikipedia.org/wiki/George-Brett)

Chapter 10

Nutrition

What Is Nutrition?

The previous chapters in this book have examined the different parts of the digestive system and the roles they play in processing the food that we eat, breaking it down into smaller parts, taking out the parts of the food that the body needs to function, and getting rid of the useless parts in the form of stool. The substances in food that supply the energy in the materials that your body needs for growth, repair, and normal function are called nutrients. The nutrients that your body requires include water, sugars (also known as carbohydrates), fats, proteins, vitamins, and minerals. The study of food and its effects on the body is known as "nutrition". People who specialize in studying nutrition are called nutritionists. They have an important role in helping planning diets for people with special needs such as those who might be lactose intolerant or diabetic.

What Is the Most Important Nutrient?

Probably the most important of all nutrients is water. Every cell in the human body requires water. The water is important for a number of chemical reactions that occur in the body. Your blood is actually mostly made up of water. When you exercise, you lose water in the form of sweat. When the sweat evaporates, it helps to cool down the body. You also lose water when breathing. You can notice this when you breathe on a window during a cold day. When you do so, the window fogs up. This fogging up is due to moisture or water particles from your breath condensing on the window. Water is also lost in the urine. Humans need to drink at least a liter or more than a quart of water a day.

What Are Calories?

One of the important things that the body needs food for is to help generate energy. Everything from running to moving your arms to even thinking requires the use of energy. Similar to a car and its need for gasoline, the body requires a source of energy to keep it going. Cells in the body take carbohydrates (sugars and starches), proteins, and fats and break them down into usable forms of energy.

One can measure the amount of energy in food by burning it. The amount of heat that is given off when food is burned can be measured. The units used to measure this are called calories. One calorie is the amount of heat need to raise the temperature of one gram of water by one degree Celsius. One thousand of these calories equal one kilocalorie. The calories referred to in diets (on the labels of food packages) or exercise programs are usually measured in kilocalories and are designated as calories spelled with a big "C". The number of Calories that are required to perform a task depends upon your age, weight, height, whether you are male or female, and how active you are (Figure 10.1). For example, a 10-year-old female who is very active may require between 1800-2200 Calories per day. In comparison, a 10-year-old male who is active may require between 2000-2600 Calories per day. In contrast, a 25-year-old female who is active may require 2400 Calories in comparison to a 25-year-old male who may require up to 3000 Calories per day. All of the foods that we eat can be broken down into energy, which translates into Calories. Some foods, particularly those that have a lot of sugars or fat in them, may have a lot of Calories in them. When one eats more food than one needs, this results in an extra number of ingested Calories. One of the ways the body tends deals with this is to store the extra Calories in the form of fat in the body. This is one of the causes of being overweight (Figure 10.2). Water, vitamins and minerals do not have Calories in them.

Activity	**Calories burned per hour**
Sleeping	55
Eating	85
Sitting	85
Standing	100
Driving	110
Walking (3 miles per hour)	280+
Bicycling	460+
Jogging (5 miles per hour)	500+
Swimming	500+
Jump rope	700+
Running	700+

Figure 10.1. Calorie burning chart for different activities - (approximate calories burned per man hour by a 150 pound person).

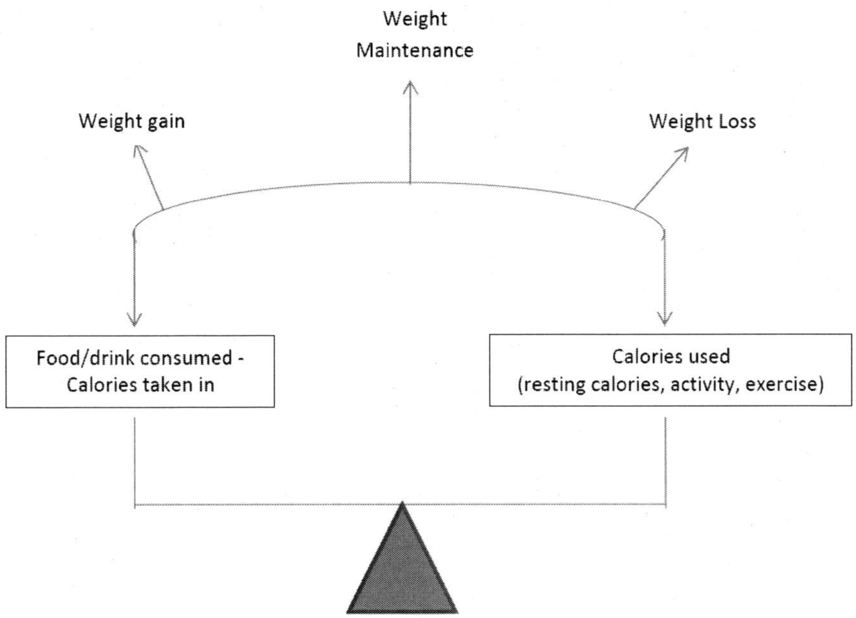

Figure 10.2. Balancing food and drink intake with Calorie use.

What Are Carbohydrates?

Carbohydrates are the main source of energy for the human body. Carbohydrates consist of sugars that are found in a variety of foods. Another type of carbohydrate is called starch. Starches are found in a variety of foods including grains, potatoes, and vegetables. Starches and carbohydrates are broken down by the digestive system into smaller sugars that can be absorbed by the intestinal tract (mainly the small intestine). The small sugars then make their way into the blood stream and travel to cells throughout the body. Another type of carbohydrate is cellulose or fiber which the body cannot digest. The cellulose ends up making up a large amount of the stool. Foods that have lots of cellulose (fiber) in them include breads, pasta (spaghetti and macaroni) and cereals.

What Are Fats?

In additional to eating fat, your body has the ability to make fat. However, there are certain types of fat that human bodies cannot make. These fats must therefore be obtained through the diet. A certain amount of fat is important in the body. It serves to protect certain organs, it insulates the body to help control temperature, and it serves a way the body can store energy. Fats are also used to make parts of cells as well as other important substance's that are needed for the body to properly function.

Fats are divided into two types, called saturated and unsaturated. The difference between these two types of fats has to do with the chemical structure of the fat. Most of the saturated fats are solid at room temperature and include things like butter and animal fat. Unsaturated fats are usually liquid at room temperature and include things like vegetable oils, which are often used in cooking. Trans fats are a type of unsaturated fat that is artificially created and put into foods to enhance flavor. Trans fats are particularly unhealthy for you.

Unfortunately, the American diet is rich in fats. Many people eat more fat than they really need. Such diets that are high in fat can cause some serious health problems including high blood pressure, heart disease, obesity and diabetes.

What Are Proteins?

Proteins are nutrients that are made up of smaller substances called amino acids. There are a total of 20 amino acids that can be combined in various ways to make up proteins. Your body uses these amino acids to build all the proteins that it needs to function. The order that the amino acids are arranged in is different and unique for each different protein. Eleven of the 20 amino acids are able to be produced by the body. The other nine amino acids need to be obtained from the foods that are eaten. These nine essential amino acids come from foods like meat, fish, milk, and eggs. Plant foods do not contain these nine amino acids by themselves. People who do not eat meat or animal products are called vegetarians. In order that vegetarians get the amino acids that they need to make proteins, they have to eat a combination of different types of plant foods, such as nuts. Proteins serve a lot of different functions in the body important for the growth of bone and skeletal muscle. Proteins aid in the digestion of food, they make substances like hormones that can be used to control sugar levels in the blood, they transport oxygen in the blood, and they are responsible for numerous chemical reactions that occur in various cells throughout the body.

What Do Vitamins Do?

Vitamins, in contrast to the carbohydrates, fats and proteins, are not a major source of energy for the body. However, the vitamins are important and are required to help regulate various processes in the body (Figure 10.3). Most vitamins are obtained from food. A few of the vitamins are made by bacteria that live in the digestive tract (recall that vitamin K can be made by bacteria in the large intestine!). If one does not have enough of a particular vitamin in her or his diet or is unable to make enough of a particular vitamin in her or his digestive tract, there can be some serious consequences. Certain types of vitamins, particularly vitamins A, D, E and K, can be stored in fatty tissues in the body. This allows the body to build up stores of these vitamins so that they can be used later when needed. Most of the other vitamins cannot be stored in fat and need to be obtained on a regular basis from the diet.

Vitamin	Sources	Function
A (retinol)	Yellow, orange, and dark green vegetables; dairy products	Important for growth of skin cells; important for night vision
D (calciferol) 	Fish oil, eggs; made by skin when exposed to sunlight; added to dairy products	Promotes bone growth; increases calcium and phosphorus absorption
E (tocopherol)	Green leafy vegetables, seeds, vegetable oil	Antioxidant; prevents cellular damage
K	Green leafy vegetables; made by bacteria that live in human intestine	Needed for normal blood clotting
B_1 (thiamine)	Whole grains, pork, legumes, milk	Normal metabolism of carbohydrates
B_2 (riboflavin) 	Dairy products, meats, vegetables, whole-grain cereal	Normal growth; part of electron transport chain, energy metabolism
Niacin	Liver, milk, whole grains, nuts, meats, legumes	Important in energy metabolism
B_6 (pyridoxine)	Whole grains, meats, vegetables	Important for amino acid metabolism
Pantothenic acid	Meats, dairy, whole grains	Needed for energy metabolism
Folic acid	Legumes, nuts, green leafy vegetables, oranges, broccoli, peas, fortified bread and cereal	Coenzyme involved in nucleic acid metabolism; prevents neural-tube defects in developing fetuses.
B_{12} (cyanocobalmin)	Meats, eggs, dairy products, enriched cereal	Coenzyme in nucleic acid metabolism maturation of red blood cells
C (ascorbic acid) 	Citrus fruits, tomatoes, red or green peppers, broccoli, cabbage, strawberries	Maintenance of cartilage and bone. Antioxidant: improves iron absorption; important for healthy gums, tissue repair and wound healing
Biotin	Legumes, vegetables, meat	Coenzyme in synthesis of fat; glycogen formation; amino acid metabolism
Choline 	Egg yolk, liver, grains, legumes	Required for phospholipids and neurotransmitters

Figure 10.3. Vitamins - Their sources and functions.

Why do people take vitamins? Next time you go to the grocery store or drugstore, you will see that there are many different types of vitamin supplements which come in the form of pills, chewables or liquids that you can buy. Some vitamin supplements just contain a certain type of vitamin while others, called multi-vitamins, contain multiple vitamins in them. These vitamin supplements can be helpful in people who do not eat well on a regular basis or whose diet is deficient or does not have enough of a particular type of vitamin. A number of vitamins are also added to certain foods, like cereals, in order to help people get the proper amount of vitamins they need in their diet each day. The body needs a certain amount of vitamins each day to make it work. It is important to realize that eating too much of a certain vitamin or taking too many vitamin supplements is not necessarily a good thing. In fact, taking too much of certain vitamins may make someone ill.

What Are Minerals?

Minerals are actually quite similar to vitamins (Figure 10.4). They do not produce energy. However, they are important to have in small amounts. Minerals can be obtained by eating various types of foods. For example, calcium is important for bone and tooth formation and growth. Calcium can be readily obtained by eating certain foods such as dairy products, certain fish, and tofu. Like vitamins, one can also buy supplements or chewables that contain many of the minerals one's body requires. If one eats a balanced diet, this is usually not necessary. Many multi-vitamins also contain several of these minerals.

Mineral	Sources	Function
Calcium	Dairy products; salmon; sardine; kale; tofu; collard greens; legumes	Bone and tooth formation; blood clotting; nerve and muscle function
Phosphorus	Dairy products; meats; poultry; grains	Bone and tooth formation; acid-base balance
Potassium	Meats; dairy products; many fruits and vegetables; grains	Acid-base balance; body water balance; nerve function

Figure 10.4. (Continued).

Mineral	Sources	Function
Chlorine	Table salt; processed foods	Acid-base balance; formation of gastric juice
Sodium	Table salt; processed foods	Acid-base balance; body water balance; nerve function
Magnesium	Whole grains; green leafy vegetables	Activation of enzymes in protein synthesis
Iron	Meats; eggs; legumes; whole grains; green leafy vegetables; dried fruit	Component of hemoglobin and of electron carriers used in energy metabolism
Fluorine	Fluoridated drinking water; tea; seafood	Maintenance of tooth structure; maintenance of bone structure
Iodine	Seafood; dairy products; iodized salt	Component of thyroid hormones
Zinc	Meats; seafood; grains	Component of certain digestive enzymes

Figure 10.4. Minerals - Their Sources and Functions.

Why Is Important to Eat a Balanced Diet?

Eating a balanced diet is important because it keeps the body operating at its very best. It would be very complicated if we had to keep track of all of the food we ate each day and try to figure out how much of each of the nutrients was in that food in order to make sure that we had enough for our bodies. In general, if we eat a balanced diet consisting of different types of foods in reasonable amounts, we can almost usually get the appropriate nutrients that we need. Experts have put together something called the food pyramid in order to help people think about trying to maintain a healthy balanced diet (Figure 10.5). The food pyramid shows that a mixture of different food types provide the necessary nutrients your body requires. Recently, the food pyramid added another important component to living a health life – exercise!

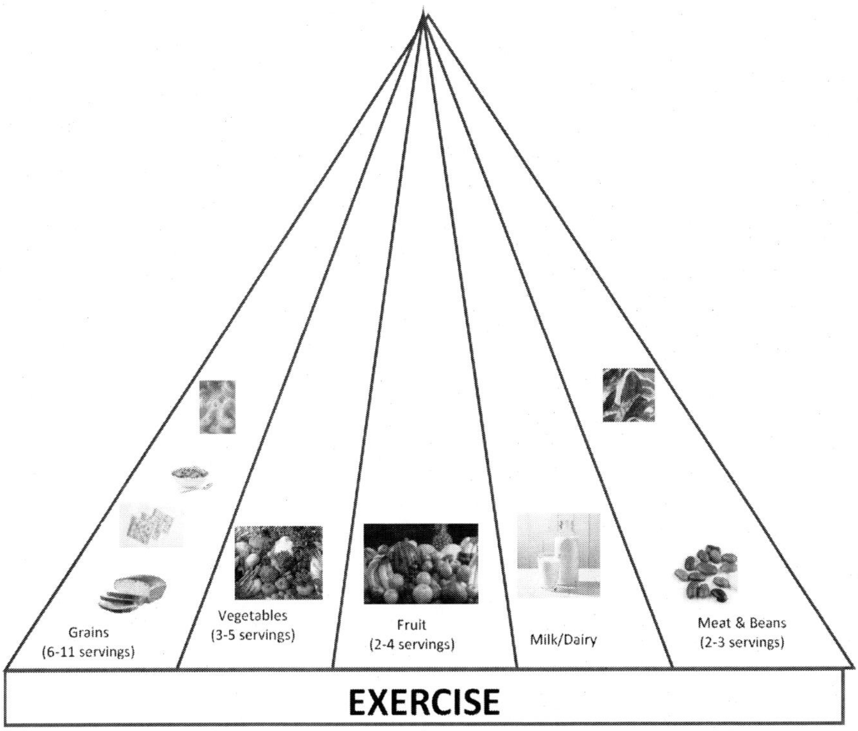

Figure 10.5. Food Pyramid - features of a balanced diet.

References

Berkowitz A. Clinical Pathophysiology Made Ridiculously Simple. Med Master Inc. Miami, FL. 2007.

Drake RL, Vogl AW, Mitchell AWM. Gray's Anatomy for Students. 3rd Edition. Elsevier Churchill Livingstone. Philadelpia, PA. 2015.

Dyer JG. Anatomy and Physiology Made Incredibly Visual. 2nd Edition. Lippincott Williams and Wilkins. Baltimore, MD. 2008.

Goldberg S. Clinical Physiology Made Ridiculously Simple. 2nd Edition. MedMaster Inc. Miami, FL. 2010.

Kapit W, Elson LM. The Anatomy Coloring Book. 4th Edition. Pearson. Cambridge, UK. 2013.

Rinzler CA. Nutrition for Dummies. 5th Edition. For Dummies. Hoboken, NJ. 2011.

Tamir E. The Human Body Made Simple. 2nd Edition. Elsevier Health Sciences. Philadelphia, PA. 2002.

Authors’ Contact Information

Richard A. Prayson, MD Med
Professor of Pathology,
Department of Pathology, L25
Cleveland Clinic
9500 Euclid Avenue
Cleveland, Ohio 44195 USA
Fax: 216-445-6967
Email: praysor@ccf.org

Nicholas F. Prayson
School of Medicine,
Case Western Reserve University
Department of Anatomic Pathology,
2109 Adelbert Road
Cleveland, Ohio 44106

Index

A

B

F

G

H

I

J

K

L

M

N

T

U

V

W

Y